CREATING SYSTEM INNOVATION

Creating System Innovation

Hans de Bruijn, Haiko van der Voort, Willemijn Dicke,
Martin de Jong & Wijnand Veeneman

Faculty of Technology, Policy and Management
Delft University of Technology, Delft, The Netherlands

A.A. BALKEMA PUBLISHERS LEIDEN / LONDON / NEW YORK / PHILADELPHIA / SINGAPORE

A Catalogue record for the book is available from the Library of Congress

Published by: A.A. Balkema Publishers, a member of Taylor & Francis Group
www.balkema.nl and www.tandf.co.uk

ISBN 90 5809 672 6

Table of contents

Preface

Many sectors face problems that call for 'system innovation.' Conventional problem solving is felt to be no longer effective. The entire underlying system of institutions, structures and values is–or needs–changing. System innovations are drastic and require the commitment of many different parties, including governments. For them, too, the question is how and to what extent system innovation can be realized or influenced. What is the government's role here? These are difficult questions: system innovation is radical and one of the lessons learned from the recent past is that the possibilities for government control are limited.

Requiring an analysis of the way system innovations come about in practice in order to draw lessons from it, the Consultative Committee of Sector Councils for Research and Development (COS) has commissioned Delft University of Technology, Faculty of Technology, Policy and Management, to carry out research into management issues in system innovation. The researchers studied three recent system innovations. They present a management perspective of system innovation based on these studies. Key questions in their studies were:

- What is the nature of the system innovation and how was it realized?
- What was the role of knowledge in the innovation?
- How did the innovators deal with social resistance?
- What was the role of public-private partnership?

The COS earlier asked the same researchers to develop an evaluation framework for knowledge-intensive public-private partnerships. The present research transcends this, placing public-private cooperation in the context of system innovation. Public-private cooperation is a possible strategy for managing system innovation, along with many others.

Harry Voorma
Chairman COS

1

Introduction: The need to change systems

1.1 SYSTEM INNOVATION

Every period faces a number of major social problems that require "system innovation". The idea behind system innovation is that regular change will not suffice to solve them. They are embedded in certain institutions, structures and values that will have to change as well. This makes the change a systemic one: it is not a matter of solving an isolated problem, but of overhauling the entire system. The isolated problem is just a component of the system. Confining oneself to solving it will, at best, plaster over the cracks and, at worst, merely aggravate the problems.

Agriculture is often mentioned as an industry needing a system innovation. It is felt that the present system of production and consumption is to blame for problems like BSE and the FMD epidemic. The system generates its own problems, which can only be solved by innovating the system as a whole.

An increase in public safety would require a system innovation. The present system can no longer guarantee safety because of problems like the limited capacity of the criminal justice system and the rise in ambiguity about norms and values,.

As we observed above, every period has its call for system innovations. Another example is the nineteen sixties, when, in the eyes of many, the North-South problems could only be solved if the underlying capitalist production system was overhauled. This resulted in another call for system innovation: a new production system would bring prosperity to poor countries.

We might also mention less dramatic and more small-scale forms of system innovation, like integrated water management, modernizing greenhouse farming areas and launching customer-oriented services in government-owned, monopolistic public transport.[1] System innovations can take place at different aggregation levels, but they always share the following aspects:

- They are comprehensive innovations, with
- a long time horizon
- requiring the efforts of many stakeholders, and
- a change of perspective and a cultural shift among these stakeholders.[2]

1.2 SYSTEM INNOVATION AND CONTROL

If system innovation is found necessary, the obvious question is how to implement it. In other words, how do we manage system innovations? The question itself implies tension. Although many writers call for radical system innovations, many others point out that the possibility of managing social processes is limited. Managing social processes is often a fuzzy, chaotic and incremental course of action, whose outcome rarely meets the expectations. What goes for social processes in general, goes for system change in particular, which is more fundamental: there is a huge gap between the desire for system change on the one hand and the possibilities of managing it on the other.

Not all system innovations result from a 'grand design'. They may develop more or less spontaneously from an unexpected angle. A well-known example is the introduction of the Internet in various sectors. Nobody planned the explosive development of Internet activities. Innovations like the rise of the Internet take place independently and decentralized, and may therefore be regarded as emergent from a central perspective. This type of system innovation, too, raises management questions. What form should the management of such emergent processes take? Can they be managed at all? Or do they just happen without the intention of creating a system innovation?

Why are the possibilities for managing system innovations so limited? Without being exhaustive, we will mention three major causes.

1. Substance: Lack of knowledge

First, there is a knowledge problem. The fundamental nature of the change implies a large number of uncertainties: what is the relation between existing problems and the existing institutions, structures and norms? Which of these institutions, structures and norms require change, and how will the change impact on the new situation? What are the financial and social costs of the change process? What type of problems will arise in the new situation? What are the costs of these problems and what will be their effects?

In many cases, there is either no knowledge at all about it, or the knowledge is contestable: different parties take different views. This makes a system innovation a risky exercise.

2. Process: Consensus is not a matter of course

This takes us to the second cause. System innovations affect the interests of many parties. In modern societies, such parties tend to operate in a network: they are interdependent, none of these parties has the power to impose its view on others. The parties involved tend to take different views about the desirability of a system innovation. Some will support it, others will be neutral, others again will seek to block the innovation. System innovation not only has a substantive component (see above). Its initiator should also be able to play the management game: who will I involve in the decision-making and how will I play the game so as to gain sufficient support for the innovation?

3. Values: Conflicting public and private values
The third characteristic is that many system changes take place on the interface of the public sector and the private sector. Agriculture and North-South problems require system innovations involving both public and private actors, each representing entirely different values. As a result, system innovations have a high potential for conflict between the public sector and the private sector, prompting the question how these should relate to each other. Is public-private partnership desirable, enabling the two sectors to cooperation in the drive for system innovation? Will such cooperation only impede progress and should a government initiate the innovation? Or is this a recipe for failure because private interests and values will then be insufficiently represented?

1.3 RESEARCH AND CASE STUDIES

System innovations tend to form the backdrop to many conflicts between public and private interests involving subjects about which insufficient knowledge is available. This causes great tension between the desirability of a system innovation and the possibilities of managing it. Nevertheless, system innovations come about in spite of this tension. What is the explanation for this? To answer this question, we identified three intriguing system innovations, we analysed how they came about and drew lessons from them for the management of system innovations. These lessons are meant in the first place for a government contemplating to initiate or facilitate a system innovation.

The research is highly explorative and the case studies are comparable only in part. The three case studies are outlined below. For each of them, we indicate both the type of system innovation.

Case study 1: Comprehensive water plan for South Florida

Since the nineteen eighties, environmentalists and the national park had been sounding the alarm: the Everglades was facing a crisis. If we did nothing, Florida's most important nature reserve would dry up. Not only would an icon of the United States disappear, but the world would lose a unique ecosystem. Alarming though the prospect might be, no action was taken for a great many years.

Two interest groups with powerful lobbies at both state and federal level blocked the discussion about rehabilitating the area: agriculture and the urban area. They cited other scientists who claimed there was no conclusive evidence that the Everglades was drying up. Moreover, what was good for a particular endangered animal species might seriously harm the ecosystem as a whole. The insoluble differences continued, and over the past twenty years 'water wars' were fought between agriculture, the city and environmentalist groups. There was a succession of court cases.

The nature reserve filed complaints against the water control authority, saying it wanted more water to preserve the crocodiles. The water control authority faced questions like: should crocodiles be saved at the expense of the drinking-water supply in Miami? Is it realistic to let farmers irrigate their sugarcane if it will cause the nature reserve to dry up?

In 1999, resistance to rehabilitation was broken. We may rightly call this a system innovation: for the first time in history, a feasible and innovative approach was adopted to the drying-up problem of the Everglades, in which parties that used to oppose each other now cooperated. Rather than fragmented plans, the parties adopted a comprehensive plan, providing an all-inclusive trade-off between the interests of the city, nature and agriculture.

The case study describes the role played by the coordinating authorities in bringing about the system innovation.

Case study 2: Privatizing British Rail

In the nineteen nineties, the British government decided on a far-reaching privatization of the rail industry. The quality of the service delivered faced widespread criticism and the costs met by the government had risen sharply. Restoring the health of the rail industry demanded huge investments, in which the government wanted to involve private parties. It also wanted to enhance the role of private parties in managing the infrastructure and providing services.

Thus, the government completely dismantled the existing institutional form, marked by an integrated rail company, British Rail, and a strong role for the trade unions. The blueprint of the alternative was marked by far-reaching privatization, both as regards the provision of transport services and the management and innovation of the infrastructure. The blueprint demanded the unravelling of many tasks, allocating them to many, often changing, parties.

To support innovation, the parties were encouraged to enter into partnerships, ending the fragmentation by sharing knowledge. The success was limited, and some of the privatization had to be reversed. Operational and strategic network management increasingly became a public responsibility, and the government had to bear more and more of the risks.

The case description shows the limitations of a system innovation initiated and controlled by a government. The blueprint offered too little room for other parties to arrange the solution and left social and technical aspects underexposed. The case study explores the mechanisms by which these technical and social aspects were found to influence the success of the blueprint.

Case study 3: The rapid rise of innovative biotech companies round Boston

A technological breakthrough in molecular biology has brought new methods for the development of drugs. The response by industry and science to this form of product and process innovation was immense. It resulted in an exponential growth of corporate activity in the private sector, much of it concentrated in a particular region. Governments created room for it by enacting a number of vital laws. These developments of products and production processes increasingly affect existing values and institutions, like ethical values (e.g. the manipulability of life), scientific values (e.g. public access to research findings) and values such as safety and public health. Several actors, including many governments on different administrative levels, try to safeguard these values. The system of values and institutions is thus being changed by an emergent, diffuse process, sidelining governments. Although there is social resistance to the technology, it can hardly get a grip on this process.

The management questions in this case study differ substantially from those in the previous case, because governments have far less control over this system innovation. What could the role of government nevertheless be? How can they safeguard public values in these rapid innovation processes? What possibilities are open to them when innovation seems to be an underhand deal between science and industry? What factors have a particular impact on these possibilities?

1.4 STRUCTURE OF THIS BOOK

We will now describe each case study in a separate chapter. Chapter 2 deals with the Everglades, chapter 3 discusses British Rail and chapter 4 describes the biotech companies in the Boston area. In each case study, we will answer the following questions:

- How did the system innovation take place? What were the main developments? What was the chronology of these developments?
- How can we qualify the type of the system innovation? What were its main characteristics?
- Given the knowledge-intensive character of the system innovation, what was the role of knowledge in the innovation process?
- Given the tension between system innovation and vested interests, did any social resistance arise? If so, how did the main actors cope with it?
- Given the fact that system innovation tends to require cooperation between public and private partners, what was the place of public-private partnership and partnering in the innovation process?

A detailed analysis and comparison, written from a management perspective, follows the case study descriptions. How can we influence the innovation process? Can we plan and manage this process? We will discuss these and other questions in chapter 5, in which we will draw management lessons from the case descriptions.

---

1. The examples are mentioned by the Dutch National Council for Agricultural Research (1999) and Vrakking (2000) respectively.
2. Dutch National Council for Agricultural Research (1999).

2

Comprehensive Everglades project: From water wars to an integral plan

2.1 INTRODUCTION

The Everglades is a unique ecosystem in southern Florida, United States, which used to be a vast area of stretches of water and grass with a rare flora and fauna. The Everglades is usually associated with the federal nature reserve in the south of Florida. However, it covers a wider area than the nature reserve alone. The 6-million-inhabitant urban area, with cities like Miami, also forms part of the Everglades. The Comprehensive Everglades Restoration Plan (CERP) covers this wider area.

Since the second half of the 20th century, attention had been sought for the fact that the vulnerable ecosystem was under threat. The Everglades were drying up. The nature reserve wanted more water for the rare plants and animals. But it was not the only player with an interest and a say in the management of the water system. Other interests had to be taken into account.

The year 1800 saw the beginning of cultivation of the area. Sugar cane is one major industry in Florida, tourism is another, its theme parks (e.g. Disney World) drawing crowds of visitors. Furthermore, Florida is the most rapidly urbanizing state in the United States, resulting in a growing urban claim on water, for both drinking water and sewerage.

The rise of agriculture and urbanization extracted vast volumes of water from the area, threatening the Everglades' unique character. Compared with 1800, for example, 98 per cent of the crocodile population has vanished[1], as have some 90 per cent of the wading birds.[2] Furthermore, urbanization is leaving less and less room for retention areas: approximately 70% less water flows through the Everglades of today than originally[3]. Lake Okeechobee has been reduced by more than half. As a result, it will easily spill over in heavy rain. The excess water is diverted to the sea by accelerated discharge, leaving it in the area for a shorter period. But the rising demand for urban and irrigation water is not the only cause of the drying up. Another major cause is the unnatural discharge regime in the Everglades.

In the last two decades of the 20th century, 'water wars' were fought between the actors representing the interests of the cities, agriculture and nature respectively. Every actor had an interest in receiving more water than the other two parties or a different water management.

For many years, the interests of these three groups seemed incompatible. To mention a few examples: the keepers of the nature reserve wanted nature conservation in the park, but received not enough water for it from the water district, or received it at the wrong times. Furthermore, the discharge regime was unnatural. If there was an excess of water, it was quickly diverted to the sea. This accelerated discharge jeopardized the survival of rare populations.

The interests of the city concern the 6 million inhabitants close to the Everglades, who need protection from floods and have to be supplied with drinking water. In addition, sugar cane is cultivated in the vicinity of the park. Sugar cane cultivation uses pesticides, which will end up in the water system. Agriculture also needs good water to irrigate the sugar cane.

Furthermore, nature conservation interests were not always explicit: an intervention in the water system may benefit one particular endangered animal species but harm another.

The initial resistance to change and differences between the parties have now been overcome. An integral water plan has been drawn up, in which all parties cooperate. Congress has approved a US$ 7.8 billion programme, with a length of 30 to 40 years. The overarching purpose of the Comprehensive Plan is 'the restoration, preservation, and protection of the south Florida ecosystem while providing for the other water-related needs of the region, including water supply and flood protection.'[4]

For the study of CERP, we have made use of various sources. In the first place, we have done desk research. Especially the Army Corps of Engineers provides the wider public with an excellent overview of the CERP process on their website[5] and numerous publications. Furthermore, we have analyzed the official documents of other actors involved in the CERP process. CERP has been object of study for many scientists, but, surprisingly, relatively few social scientists. we have made use of the work of influential scholars such as Davis & Ogden[6]; John C. Ogden[7]; Gunderson[8] and the study by van Eeten and Roe[9].

On the basis of the overview of the actors involved (see above), we have selected five interviewees. The first interviewee was part of the academic community that made the first initial steps in a process that would later result in CERP. He is chair of a department of environmental studies. It is confirmed by other knowledgeable people that this person played a key role in the process. Recently, he is not as much involved in the process as he used to, which enables a distant view. The second interviewee is part of the academic network too in the field of Biological Resources. He has continued to work within CERP from the beginning until this date. He is an expert in integrated modeling, a feature that has proven to be of crucial importance in this process. Furthermore, two persons from the SFWMD are interviewed, both the Chief Scientist and the Lead Ecologist. These persons too, were part of the CERP community from the early beginning and established the

Science Strategy, a vital element in the CERP process. Finally we have interviewed the senior officer at ACE (Chief of the Recover Branch), who is the responsible officer for CERP within the US Corps of Engineers. All interviews were conducted in 2002.

2.2 A BRIEF CHRONOLOGY

Resistance to rehabilitation

In the mid nineteen eighties, a number of environmentalist groups realized that the Everglades was facing a crisis. Populations of rare animal species (e.g. birds, tigers and crocodiles) were threatened with extinction and the Everglades was in danger of drying up even further. A lobby for the 'Restoration of the Everglades' was formed, which gave rise to a counter-lobby. The sugar farmers and the Native Americans were fiercely opposed to rehabilitation of the area. The farmers feared they would have to leave the area. They had traditionally had a strong lobby at the federal level and with Mr Jeb Bush, Florida's governor.

The Native Americans had another reason for opposing rehabilitation. They had full autonomy within their territory and feared that they would lose control if the nature conservation became the first priority.

The state of Florida also had reasons to oppose the plan. Water management is a state affair. Conservation of nature, however, is a federal responsibility. Hence this would mean loss of authority: 'Water belongs to Florida, not to the US. There was a fear for federal control.'[10] Florida is the most rapidly urbanizing state of the US. It was feared that urban water management interests would be sacrificed to environmental interests.

Until the nineteen nineties, numerous court cases were brought. Environmentalist groups sued farmers about irrigation practices and the amount of pesticides in the water, discharged from the land into the ecosystem. Native American tribes sued the South Florida Water Management District (SFWMD) about its policy. The national park sued both farmers and the state. Cities sued the SFWMD because they did not receive enough drinking water. The water problem seemed insoluble.

The role of expertise

In the mid nineteen nineties, 21 separate environmentalist groups formed a coalition for the Restoration of the Everglades, a powerful political lobby because of its size, but also because the political tide had started to turn.[11]

During the same period, a group of biologists, hydrologists and ecologists from several public organizations conceived the plan to enhance their cooperation. They were employed by three organizations: the South Florida Water Management District, a kind of water control authority at state level;

the Army Corps of Engineers, responsible for the infrastructure, a federal organization; and the biologists, hydrologists and ecologists attached to several universities. Although these organizations did not cooperate, their biologists, hydrologists and ecologists proved to speak the same language.

Fed up with the managers disregarding their new insights, they formed a 'Science Strategy', frequently holding workshops and exchanging models from the individual disciplines (i.e. biology, ecology and hydrology). These meetings of scientists were highly important, because this was the first time ever that scientific consensus was reached about the main 'stressors' of the Everglades[12]. This was remarkable because, until then, there had been discord about the most basic issues, like the question whether the Everglades was getting wetter or drier. Central to the report[13] of the Science Strategy Group was: 'let's not worry about the things we do not know. Let's organize the knowledge we have in such a way that it is more effective.'[14] The group translated scientific insights into managerial tools: what can a manager do about these 'stressors' in the system? With these tools the Strategy Group informed the managers of the responsible agencies: 'Look, here are the key stressors. On all these stressors there is scientific consensus'[15]

Significantly, 'multi-modelling' was made possible for the first time. Models of biologists, hydrologists and ecologists were integrated. For the first time, questions could be answered like: What does a certain water level mean for endangered animal species A?

This report from the 'scientists' caught the federal government's attention, prompting it to conduct further research on whether the Everglades indeed faced a crisis, and if so, to find the possible solution. The Clinton administration proved receptive to the issue. The government was eager to use the case to demonstrate its environmental awareness. Furthermore, the Everglades was held in high regard by many Americans. The Strategy group used this fact in a clever way, by consistently speaking of 'America's Everglades'[16]. It worked. The region was viewed as national heritage and saving it an undertaking of national significance. It was thus accorded high priority at the federal government level.

Yet the ramifications of the report again led to much opposition, not only from the sugarcane farmers but also from the state of Florida. These players were in sore need of more water to supply some of the fastest-growing cities in the United States. Their fear was that improvement of the Everglades would only concern 'restoration of the eco-system'. If the Everglades was going to be 'restored', what would happen with the growing demand for drinking water supply and agricultural use? At the federal level the decision was made that more research had to be done.

The role of scientists is considered to be absolutely crucial for the whole process: 'science created the restoration, not the managers'.[17]

'Enlarging the water pie': resistance broken

Subsequently, the Restoration Study (Restudy) was launched in 1993[18]. What was crucial was the setting up of a Governor's Commission, headed by Richard A. Pettigrew, bringing together 16 stakeholders. Its mandate: 'to form a unanimous opinion on how the water problem should be tackled within 16 months.'[19]

The Restudy team, comprising hydrologists, biologists and ecologists, operated in parallel with this commission. The Restudy team set to work developing models and translating them into language that would be understandable for non-scientists. They went 'back and forward' between the Restudy group and the Governor's Commission in 'numerous meetings'[20].

This group was skilful in reframing issues and choosing the right metaphors. For instance, they did no longer use the wording 'Restoration', since this concept was emotionally charged. It was equated with ecological restoration. In order to get the sugar cane farmers, the tribes and the cities aboard again, they chose for 'Enhancement'.[21]

A major milepost was the Restudy team's finding that the water budget could be doubled: 'enlarging the pie'. An integral water plan could be developed that forced nobody to give up water. There would be more water for nature, for the city and for agriculture. The use of new technology and better timing of the discharge regime would allow the water budget to be doubled. Thus, nobody would be forced to sacrifice water. An integral water plan would make everybody a winner.

The doubling of the water budget would be achieved by improving the timing of the drainage regime. Moreover, new technologies were envisaged. Since Aquifer Storage and Recovery (ASR) is key in doubling the water budget, we will explain this technology very briefly.

ASR refers to the process of recharge and storage of water in an aquifer system during times when water is plentiful (typically during the wet season in south Florida) and recovery of the stored water during times when it is needed (ACE, 2003c). ASR can function in the manner of a traditional surface water reservoir; however, aquifer storage eliminates evaporative losses, and the requirement to convert large land areas into reservoirs. It is thought to increase availability of large volumes of water during severe, multi-year droughts to augment deficient surface water supplies. CERP includes 333 ASR wells with a total capacity of 1.6 billions gallons per day. Much of the 'new water' in CERP is derived from storing excess water that was previously discharged to the ocean.[22]

The Restudy members submitted this view (i.e. doubling the water budget is possible) to the Governor's Commission. They also made clear to the farmers that they, too, had a water problem. Without an integral water plan, they would already lack water in the medium term.

Everybody accepted the plan. In 1999, CERP was presented to Congress.

The US$ 7.8 billion programme was approved in 2000. Half of it would be funded by the federal level, the other half by the state of Florida. The plan's implemenation period was set at thirty to forty years.

The present state of affairs

When the plan was submitted, the proposers reluctantly admitted that the technologies described were still uncertain. In the Summer of 2002, this and other technologies were heavily attacked in the public debate: 'the plan relies on four highly speculative technology gambles that account for nearly half its price tag.'[23] The proposers used terms like 'pilot projects' and 'adaptive management' to deal with this uncertainty. There were also a great many 'open ends'. The details had not been worked out, and the proposers lacked vision about the plan's implementation. Critics said 'open ends' was the 'euphemism of the century', alleging that the plan ignored major issues like water quality, rising sea levels, maintenance costs and doubts about the feasibility of the technological solutions.

By this time, there were frictions within the coalition. Apparently, doubling the water budget was less easy than expected. The technological solutions turned out to cost more and some of them were unworkable. Particularly the park felt that it would not receive enough water (quantity), that it would receive it too late (timing) and that it will not be clean enough (quality).

Furthermore, there were clashes between the federal level and the state level, between agriculture and environmentalists, and between the 'single species' and the 'ecosystem' perspectives (i.e. the dilemma that the ecosystem as a whole will suffer by preserving one protected animal species).

Currently, a vexed question is whether the guarantees for environmental investments should be laid down in law. The Army Corps of Engineers is now developing guidelines containing process proposals: an external 'review commission' will be set up, and interim reports have to be submitted, comparing the state of affairs with performance indicators. The struggle is about what should be laid down in the law and what should not. The park and the environmentalists want to lay down as much as possible, whereas he state of Florida wants to leave open as much as possible.

Here, we see the tension between uncertainty on the one hand and the drive for 'on time, on budget' on the other. SFWMD would like to agree new rules when new trade-offs are made, whereas the environmentalists would like to see the guarantees written in stone.

Structure of this chapter

This chronological account will be followed by a critical discussion of the system innovation. Although we can call this integral plan a successful system innovation in many respects, the success also has its downside. This observation will feature in all sections of this chapter. We will first describe the nature of the system innovation and the role the coordinating governments played in it. This case study will focus on the fact that the system innovation could be achieved because of the way the governments dealt with social resistance. This is the subject of section 3. Section 4 discusses the role of expertise. In the final section, we will analyse the way of 'partnering.'

2.3 THE TYPE OF SYSTEM INNOVATION

After twenty years of water wars between the Everglades' stakeholders, a breakthrough has now been reached that will help the fight against the drying up of the Everglades. Local and national governments have together formulated the Comprehensive Everglades Restoration Plan (CERP), comprising an integral water plan for the whole area. We regard this plan as a large-scale system innovation, not only because of the project's magnitude (7.8 billion dollars) or duration (30-40 years) but especially because of its long-term and substantial consequences:

- For the first time in history, a workable approach to the desiccation problem of the Everglades has now been adopted.
- There is cooperation between parties that used to oppose each other.
- A comprehensive plan has superseded fragmented plans. There is now an integral trade-off between the interests of the city, nature and agriculture.
- The plan is based on an integrated model made by hydrologists and ecologists for the total area.
- There are new technological solutions for the water problem.
- The plan has resulted in a more natural discharge regime, considered to be more sustainable than the regime of the past 40 years.

From emergent to planned system innovation

The case study description shows that the innovation started as an emergent process. Bypassing the institutional boundaries, ecologists, biologists and hydrologists initiated the organization of workshops. In them, scientific consensus was reached for the first time about the most basic questions like: are the Everglades drying up or is the area getting wetter? Following this phase, it was asked what human interventions were the system's greatest 'stressors'.

The ecologists, biologists and hydrologists had not been commissioned by their managers or the ministry to organize these workshops. Their action developed bottom-up out of common frustration. They had in common their disappointment at the managers of both the Army Corps of Engineers, the federal responsible organization and the South Florida Water Management District, the organization at state level, taking insufficient action to save the Everglades. One of the reasons for the lack of resolve was the fragmentation of scientific insights and the poor translation of those scientific insights into managerial practice. All of them felt the need to translate scientific insights into '*management tools*'.

'Science Strategy', the report produced by this group of scientists in 1995, prompted the federal administration to commission research into the condition of the Everglades. We see the system innovation being planned from the time of the enactment by Congress of the Water Resource Development Act (1996). The SFWMD and the Army Corps of Engineers were the coordinating actors.

The conclusion is that the system innovation began as an emergent process. The problem identified bottom-up managed to draw the attention of the federal level. From then on, the process was planned.

2.4 THE ROLE OF KNOWLEDGE

Scientific consensus created

In the mid nineteen nineties, there was no scientific consensus about the most basic and most pressing questions regarding the Everglades. Some thought that the Everglades had become dryer, whereas others thought that the Everglades had become wetter. [24] The 'Science Strategy' (1995), in which ecologists, biologists and hydrologists reached consensus both about the state of the Everglades and about the main assignment for future research, was a breakthrough for the CERP process. For the first time, scientists were able to present unanimous insights to the managers.

An important comment should be made on this 'scientific consensus', because the three disciplines referred to represented just part of the scientific disciplines involved (e.g. the group does not comprise geologists, sociologists, economists, etc.). Furthermore, the views of professional engineers were not represented. This narrow basis, however, also had its advantages. It was easier to reach consensus between these disciplines than in a wider group.

A second comment is that, in fact, the Science Strategy report presented rather simple, broad outlines. The advantage of these 'outlines' was that they were easier to communicate to the managers.

A more subtle approach was lacking in two respects. First, it was a rather narrow basis. Second, the insights were still at the early stage of broad outlines. These two elements accelerated the process: consensus could be reached soon and the managers had relatively simple dilemmas and solutions submitted to them.

Convergence of knowledge crucial

A critical factor in the system innovation was 'multi-modelling'[25]. Scientists from three different disciplines together developed a model that integrated the views from ecology, biology and hydrology.

When these scientists formed a joint workgroup, crucial questions were answered that were directly relevant to the managers. Before the workgroup was formed, scientists had indeed scored results in the individual disciplines, but these insights were too fragmented and too specialist for the managers. Having a model that integrated the three disciplines, they could now answer questions like: "What interventions have the most disrupting effect on the Everglades?" Or 'What does a particular water level mean for a particular bird species?'

The integrality of the knowledge contributed to the managers sharing the urgency of the problem with the scientists.

Interface between science and management

Right from the start, the scientists sought knowledge that was both accessible and useful to managers. From the outset, they focused on the interface between science and management. They simplified the models and used attractive visualizations 'allowing any lay person to understand them'. Videos were made to visualize the abstract models.

The ecologists, biologists and hydrologists not only analysed the problem, but also developed 'management tools' to address the problems in the Everglades.

Coupling political priorities

Apart from focusing on the managers, the scientists were always conscious of the political arena in which they operated. Interviews showed that the scientists realized quite well that the rehabilitation plan could only succeed if the farmers and the state of Florida gave their cooperation. Everything was intended to create a 'win-win' situation, as was clear from a reformulation of the original plan (i.e. restoration) into 'Enlarging the water pie'.

All in all, we do not go as far to agree with one of the interviewees, that 'Science created the restoration, not the managers'[26], but it is clear from this description that the creation of consensus among scientists had been crucial in the process.

2.5 SOCIAL RESISTANCE

The chronological description (section 2.2) shows a growing awareness since the nineteen eighties that the Everglades was facing a crisis. Both environmentalists and the national park were unanimously fighting for rehabilitation. Yet, no system innovation was implemented. The reasons were that, until the end of the nineteen nineties, there was no *sense of urgency* on the federal level, and that many actors opposed rehabilitation, for which everyone had their reasons:

- The *farmers* grew sugar cane, a crop that needs a lot of water. Some people said that sugar cane was not a sustainable activity in an area lacking water. Sugar farmers feared that, in the long term, they would no longer be allowed to farm in this area and opposed any form of rehabilitation.
- *Native Americans* have full autonomy within their territory. They feared that they would lose control if conservation was put first.
- The state of *Florida* had another reason to oppose the plan. Water management is a state affair. Conservation, however, is a federal responsibility. Furthermore, the state also looks after the interests of the urban areas. Florida is the fastest urbanizing state in the US. It was feared that interests of urban water management would be sacrificed to environmental interests.

Below, we will analyse how this resistance was broken and what role the coordinating actors played. Two aspects prove to be crucial: first, the *framing* of problems; second, the arrangement of the process, particularly the relation between *stakeholders* and professional engineers/scientists.

From single-purpose to multi-purpose

When, in the nineteen eighties, the national park and environmentalists submitted the plans for rehabilitation, rehabilitation was presented as an aim in itself. This heightened the differences between the parties. After all, if 'nature' was an aim in itself, a claim by the urban area or by agriculture could never be justified, because such claims would always be made at the expense of the water available for the eco-system.

The resistance by the farmers and the state was not broken until rehabilitation was no longer presented as *single-purpose* ('rehabilitation of the ecosystem'), but as *multi-purpose* ('increasing the water budget for everybody'). This meant that new technological possibilities and a different way of management would make more water available for everybody. Not only the ecosystem would gain by rehabilitation, but there would also be more water to divide among all parties involved.

The reformulation of the project offered gain, for the first time in twenty years, to farmers, cities and the national Park by adopting a cooperative stance in the rehabilitation process.

Parallel tracks of scientists and stakeholders

After many workshops, the 'scientists' of both the South Florida Water Management District (i.e. the state) and of the Army Corps of Engineers (i.e. the federal level) reached consensus in 1996 as to what the main problems were for the Everglades and in what direction a solution should be sought. What remained now was to convince the *stakeholders*.

A parallel process was arranged for this purpose. On the one hand, the scientists would continue to develop directions towards a solution. This was the 'Restudy' track. At the same time, the main stakeholders were brought together (i.e. the Governor's Commission). The group of scientists kept this commission informed about the pros and cons of certain solutions proposed by the stakeholders and about whether or not particular wishes could be realized.

Although there was no official coupling between the two tracks, the dynamic between the two groups was very productive. In fact, the scientists indicated the frameworks within which the *stakeholders* could reach a solution. In their turn, the *stakeholders'* preferences steered the way the Restudy group emphasized certain issues. Because of this loose coupling, the group of stakeholders unanimously adopted the Restudy group report.

Incentives

This was a highly expensive project. Most parties had no financial incentives to participate. However, the negative financial incentives were substantially mitigated because the Everglades was high on the federal agenda. If Congress approved CERP, US$ 7.8 billion would be available. The federal level would pay for half of this project, the state of Florida for the other half. Without CERP, there would be no contribution from the federal level.

There were positive professional incentives for scientists to participate in this project. If it was implemented, the ecologists, biologists and hydrologists involved would have the opportunity to develop a refined multi-model bringing substantial innovation for all individual disciplines. The professional incentives therefore stimulated the realization of the project. The financial incentives were not an obstacle because of the 'matching' principle of federal funds and funding by the state.

If all aims of the integral Everglades project were achieved, the sugar farmers, the cities, the national park as well as the Native Americans would all benefit. Some of these advantages were visible, others were not. Particularly the advantages for the ecosystem were difficult to measure. First, because they would only become manifest in the long term. Second, because those advantages were difficult to establish. Seemingly simple questions like 'Has the population of a particular endangered animal species increased or decreased?' were difficult to answer because counting these animals was very complex.

2.6 PPS AND PARTNERING

Public-public partnering, with stakeholder participation

CERP is an entirely public affair. Although the stakeholders – some of them private – are involved in the planning, there is always a loose coupling between advice and the eventual decision. In the Governor's Commission referred to, for example, the stakeholders have the right to give advice. The Restudy team, however, has no official obligation to follow their advice.

An American phenomenon plays a part here. To prevent certain groupings in society from having disproportionate access to decisions of public organizations, non-public officials are not allowed to hold seats on federal and state commissions. The law provides that only civil servants may be members of these commissions.

One reason for this public-public cooperation is that substantial resources are available for the rehabilitation project. It surprised some of the parties involved that Congress was willing to spend so many dollars for 30 years on a water problem that could have been regarded as a local problem. The reason for the federal involvement is that the park is regarded as a national heritage, as an icon of the United States. Furthermore, both Al Gore (during the Clinton administration) and Jeb Bush, the governor of Florida, used the project to present themselves as Green administrators. Ample resources could be claimed because the Everglades is a major national symbol for Americans. This is why there was no incentive for private involvement. In other words, one disadvantage of public-public cooperation was the lack of *incentives* for creativity, realism and budget control.

An advantage of this public-public cooperation was that, at the start of the project, there was little organizational complexity, because ACE and SFWMD coordinated the whole project. This led to considerable momentum in the early stages of the process. Later, however, fragmentation occurred. Stakeholders that had been kept out of the process until then started to make themselves heard, such as the environmentalists committed to the conservation of a certain animal species. Supported by federal laws protecting endangered animal species, they intervened successfully. As the process proceeded, the initial momentum gradually declined.

Reasons for cooperation

The project was, and is, being managed by the South Florida Water Management District and the Army Corps of Engineers together. The reason for this cooperation is both financial and substantive. The financial reason is that CERP is funded 50% by the state and 50% by the federal level. The two organizations each represent a layer of the funding.

Further, the Army Corps of Engineers is responsible for the infrastructural projects that have to be carried out for water management. CERP comprises many infrastructural projects. Irreverently, it is sometimes referred to as a 'plumbing project'. ACE is *the* organization to contribute expertise. SFWMD is responsible for water quality and water quantity in southern Florida.

---

1. US Corps of Engineers, 2003a
2. Ogden, 1999
3. US Corps of Engineers and SFWMD, 1999a
4. Water Resources Development Act, 2000: section 601 (h)
5. Army Corps of Engineers, 2003a
6. Davis & Ogden, 1994
7. Ogden, 1999
8. Gunderson, 1999
9. Van Eeten & Roe, 2002
10. Interview with Ogden, 2002
11. Davis & Ogden, 1994
12. Interview with Davis, 2002
13. Davis & Ogden, 1994
14. Interview with Ogden, 2002
15. Interview with Ogden, 2002
16. Interview with Appelbaum, 2002
17. Interview with Ogden, 2002
18. Army Corps of Engineers, 2003b
19. South Florida Ecosystem Restoration Task Force, 1999
20. Interview with Davis, 2002
21. Interview with Ogden, 2002
22. Army Corps of Engineers, 2003c:5
23. Washington Post, 23 June 2002
24. Interview with Gunderson, 2002
25. Interview with D. DeAngelis, 2002
26. Interview with Ogden, 2002

3

The rail revolution: Institutional refinement of a revolution

3.1 INTRODUCTION

There are times when governments have the means to initiate a system innovation themselves. This seemed to be so in the following case study. In the nineteen nineties, the British government decided to rearrange the rail industry. It charted a dramatically new course on the border between market and government. The rail revolution was a fact.

This chapter will attempt to picture the shaping of the rail revolution, its implementation and its effects. What was the idea behind the design of the rail revolution? In what context, both technical and social, was the revolution implemented? What were the consequences of the revolution, such as those regarding the intended results?

3.2 A BRIEF CHRONOLOGY

3.2.1 THE RAIL INDUSTRY BEFORE THE REVOLUTION

The rail industry in the United Kingdom has built up a long history since trains started running here in the early years of the 19th century. The industry flourished in private hands and went through difficult years after the First World War. In 1921, the government intervened in the industry, enforcing consolidation. The existing 123 companies were merged into four large companies. Round the Second World War, the industry again faced problems and the government intervened once more. On 1 January 1948, under Atlee´s Labour government, private operation of the railways was halted and the whole of the industry passed into government hands. This was the beginning of British Rail. The movement was similar to that in the Netherlands. Although consolidation had taken place earlier in the Netherlands, the government completed the integration of rail carriers after the First World War and established Dutch Railways, whose sole shareholder was the State of the Netherlands.

As early as the nineteen sixties, doubts arose about the quality and efficiency of the service delivered. However, the Beeching Commission[1] did not intervene in British Rail's organization, but in its service provision, which led to the closure of over 2,000 small stations and 3,000 kilometres of railway line. The originally Conservative plans, launched by the MacMillan government, were implemented under Harold Wilson´s Labour government. However, British Rail retained its monopoly on the British rail market.

In the following years, British Rail went through a turbulent patch. Frequent strikes interrupted services, like the one in July 1989, which caused the usual problems, not least in the metropolitan area. The quality of the infrastructure was lagging behind, and frequent accidents added to doubts about the quality of British Rail's internal management.

As these doubts continued, the Thatcher government decided on a large-scale intervention in the industry's organization. In 1993, the Major government published the green *paper New opportunities for the railways: the privatisation of British Railways.* It contained proposals for a radical restructuring of the organization of the rail industry, based on two guiding prin-ciples: liberalization by franchises and privatization by splitting up the industry[2].

Leading economists were involved in working out the eventual form of privatization and liberalization. They opted for a horizontal as well as a vertical splitting up of British Rail. The bottom layer comprised the infrastructure: rail systems and stations. These were transferred to Railtrack, a private monopolist organization that took over the whole of the infrastructure, funded by a share issue. There was no horizontal splitting up on this layer. Railtrack also became responsible for maintenance and management of the infrastructure. It contracted out the work to private contractors, many of them originating from British Rail[3].

The intermediate layer comprised vehicles. British Rail's vehicles were sold to three rolling-stock companies (ROSCOs), which leased the vehicles to the operators. This was expected to simplify these operators´ entry into the market. There would be competition here, and a vertical splitting up.

The top layer comprised operations. They were issued in franchises to several private carriers, the train-operating companies (TOCs). The process started by franchising the attractive intercity services from and to London. Then, the remaining regional services were put out to tender. The TOCs could tender for franchises comprising "train paths"[4] for a certain period. This allowed different carriers to make a regulated use of the same track, such as an intercity-service carrier and a slow-train carrier. The government selected carriers by means of a “beauty contest”, run by the Office of Passenger Rail Franchising (OPRAF). OPRAF was incorporated into the Strategic Rail Authority (SRA) in late 1999. The SRA was also involved in developing the infrastructure.

The government still had a major say in determining the services provided by choosing train paths. Apart from the SRA, there were two other, important parties on the government side. The first of these is the Rail Regulator (RR), which oversees market relations in the industry. Another important role is reserved for the Health and Safety Commission and Executive (HSC/E), which monitors safety in the industry.

3.2.2 THE RAIL INDUSTRY AFTER THE REVOLUTION

Revolution and resistance

In many respects, the restructuring of the industry by liberalization and privatization was a revolution. The privatization of practically the whole of

British Rail meant that the organization had to be split up into a large number of components. For most of the staff, it meant a changeover to the private sector, with all the uncertainties it entailed, after years of solid embedment in the public sector. Privatization had to make competition possible. This caused a shift in the position of the British government, which had always managed the railways through the British Railways Board on very generic elements. The Board was hardly involved in determining service provision and its implementation. However, the form of competition it had chosen, viz. by tendering out franchises, forced the government to define the franchises and choose the most attractive offer. To guarantee safety, it intervened more directly in the internal management. Passengers witnessed an increase in the number of rail-transport providers from just one to dozens. In the process, carriers began to address passengers as "customers" rather than "users".

Already at the start of the privatization and liberalization, considerable resistance to privatization emerged, especially on the part of British Rail staff. This resistance grew over the years, when the quality of service failed to improve substantially. Although carriers improved the comfort of the carriages, the dense networks round the big cities continued to be plagued by breakdowns and delays. The public increasingly blamed privatization. This criticism peaked immediately after two major rail crashes, the one near Paddington in 1999 and the one in Hatfield in 2000.

Many of those involved judged that the problems in the industry were due to the fragmentation that had followed privatization. A majority of the British population supported the renationalization of the railways[5]. The Paddington committee of inquiry headed by Lord Cullen[6] concluded that the industry lacked a clear leader in the area of safety.

Fragmentation and contractualization

Many saw fragmentation in the industry as one of the aspects compromising safety. Coordination between the different layers, infrastructure, vehicles and operations had always been an internal optimization problem within British Rail. The splitting up, however, made it an interorganizational problem, with a different way of handling interruptions. In their perspective, that was a more difficult assignment[7].

Both vertical and horizontal coordination were necessary. Franchises had no exclusive use of tracks. In addition, it was decided to have different franchises for intercity trains and slow trains. Different operators shared the use of most tracks, several carriers having been allocated different train paths on the same track. Coordination problems between these carriers occur, for example when one of the carriers suffers a delay.

To manage this interorganizational coordination, it was decided to introduce far-reaching contractualization of relations between parties, in particular between Railtrack and the carriers. Deviations from the train paths as laid down in the franchise have been extensively monetarized. For example, the causes of delays are traced and causers have to pay Railtrack a fine. Railtrack has to pay fines to the carriers if it deviates from the train paths laid down in the franchise, either because of delays of other carriers or maintenance that deviates from the room laid down in the franchise.

The contracts contain a detailed system of bonuses and penalties for operational performance, enhancing the focus on operational availability and reducing the focus on innovation. Innovation brings immediate costs, whereas the long-term proceeds are unclear, since they depend on the expected breakdowns of the existing infrastructure.

Constant failure

In some cases, the contractualization led to excesses. Stagecoach Rail director Graham Eccles8 gives an example of this. Virgin Trains, whose shares are held by Stagecoach, receives considerable compensation for work on the West Coast Main Line (WCML) overrunning its schedule. Virgin Trains is the main user and was awarded a franchise guaranteeing the output of the track during the adjustment operation. The contract for the upgrade left little room for decommissioning for and overruns of work on the tracks. Virgin received substantial reimbursements for uncontractual decommissioning of the WCML, which contributed to the project being over budget.

In addition, fragmentation will reduce the focus on a joint problem approach, as parties will focus on their own responsibility[9]. Parties are well aware of their own tasks within the operation and confine themselves to them. In an interview, Graham Eccles[10], Stagecoach's Director of Rail, indicated that the company is concentrating more and more on what it is good at: running trains and buses. Its interest in investment and development at the start of the privatization and liberalization has waned.

Although contractualization tried to harness fragmentation, it caused other problems. In an interview[11], David Thomas, Director of Corporate Finance of the SRA, referred to the problems triggered by replacing a large number of trains of the Mark I type in the region south of London. In the process, a vehicle maker eventually built a large number of vehicles and supplied them to ROSCOs for a regional carrier. However, the vehicles demanded too much electricity of the existing infrastructure, owned by Railtrack.

Questions were also being asked about track maintenance. Some accidents (e.g. Hatfield, Potters Bar) were associated with failing maintenance. Especially as regards the Potters Bar crash, the finger was pointed at fragmentation, enhanced by subcontracting. This widened the distance between the maintenance firm and the carrier.

Ian McAllister, the present chair of Network Rail, which succeeded Railtrack in October 2002, regards fragmentation between parties as a major problem. In his view, Network Rail should bridge the "damaging rift between train operators and track engineers"[12]. Graham Eccles also feels that close coordination between parties is necessary.

From 1999, the system innovation was under great pressure. Apart from the doubts about safety, Railtrack proved unable to sort out its internal management. Unable to earn a profit, the company, listed on the stock exchange, frequently asked the government for substantial extra contributions. In 2002, this eventually led to the sale of most of the infrastructure and the handing over of tasks and employees to Network Rail, a "not-for-dividend" enterprise.

Throughout the introduction of the system innovation, different forms of public-private partnership were used. In the United Kingdom, a distinction is made between "track maintenance, renewal and enhancement". The first two were made the responsibility of Railtrack–of Network Rail later–and should be carried out with their own resources; this is standard maintenance and the replacement of existing tracks. Enhancements are real changes to tracks. For these, the SRA is responsible. For *enhancements*, frequent use is made of "Special Purpose Vehicles" (SPVs). They contain the institutional contracts between the parties involved in the improvement. We will discuss the role of the SPVs in the system innovation in more detail in section 3.6.

3.3 THE TYPE OF SYSEM INNOVATION

The system that is the subject of the innovation described here is the British rail industry. It provides transport services for goods and passengers. The system can be subdivided into a technical and an institutional part. The revolution particularly involved the institutional part, whose effects on the technical part we will also set out here.

The technical system: low failure tolerance and many couplings

Providing services in the system is an active process: the system functions only if relatively great efforts are made, which makes the system inherently instable. The system has a low failure tolerance, because system failure is highly visible and directly affects large groups of stakeholders, notably passengers. Initially, a failure is concentrated in larger events, more than in road traffic, for example. One single derailment will easily be a disaster, whereas one single car leaving the road is an incident.

In addition, a failure rapidly expands because of the many couplings in the system[13]. It has many interdependent subsystems, both horizontal and vertical. First, the system has relatively hard vertical couplings between the layers mentioned above. In fact, these layers are coupled capacity systems: infrastructure, vehicles and staff. Via these couplings, failure in one capacity system will trigger failure of the other capacity systems. If train drivers are on strike, no trains will run and the infrastructure remains empty. If there is a lack of rolling stock, quickly deploying staff usefully elsewhere will be impossible and the infrastructure will remain underused. When wires break, both staff and rolling stock will be at a standstill.

In addition, the system has horizontal couplings in the form of trains delaying each other. On the tracks, there are limited possibilities for overtaking, and a delayed train will hold up other trains on busy tracks. This is a hard coupling. If there are several tracks, used less intensively, this coupling will be less decisive. There are also soft couplings in the connections between different train services at railway stations. Although these couplings are soft from a technical perspective, they determine the quality of the system. They also cause failures to spread through the system; minor disruptions cause major delays.

The old institutional design

For decades, instability and low failure tolerance have led to a strong technically- oriented approach in rail transport. The strong couplings are both the cause of the instability and the basis for an institutional form comprising the couplings: a single organization for infrastructure, vehicles and services. In Britain, this was British Rail and in the Netherlands it was NV Dutch Railways. These integration-oriented organizations chose a strategy oriented towards organizational planning and technical redundancy to limit failure. This approach is also dictated by the limited possibilities of a more flexibility-oriented approach. Investments have a long lead-time and a long return time: infrastructure has a life of about one hundred years, vehicles last some thirty years. There is little flexibility in the system.

The institutional form not only results from the many technical couplings in the system but also from direct public intervention. Both in the Netherlands and in Britain, full integration was initiated by the national government; after the First World War in the Netherlands, after the Second World War in Britain.

This institutional integration produced a monopolist with a strong internal focus, in which passengers were an assignment rather than customers. In the nineteen eighties, this was increasingly regarded as problematic, not only in Britain, but in the rest of Europe. The European Union saw the existence of the national monopolists as market protection and advocated a separation of infrastructure on the one hand and service provision, including vehicles, on the other. This was to make the entry of new carriers possible.

The new institutional design

Consensus democracies like the Netherlands opted for a limited introduction and a slow process. The British two-party system, however, opted for a revolution, with the Conservatives opposing the old British Rail as an exponent of Labour thinking. British Rail was soon split up into different components and a new institutional context was designed on the drawing board. Frequent failures of the system clearly warranted the dramatic change. British Rail being a public organization gave the Conservative government a grip, making revolutionary reorganization possible.

The system innovation was clearly planned: the desired final situation was designed and introduced in a relatively short time, mostly by policy makers and advisers. Essentially, the different layers were placed with different private enterprises, most of which had been, or were, listed on the stock exchange. The couplings between these enterprises were covered by bilateral contracts rather than by an umbrella organization.

A foreign body in the new institutional context was franchising. The framers of the institutional design embraced the idea that there could be no unregulated entry to the railways; the couplings required some coordination. To regulate rail entry, a public body was set up to assign train paths (see footnote 4) to a single carrier for a certain period.

The new institutional context occasioned a dramatic shift in the role of government and industry, causing upheaval in the value systems. One example is membership of the two trade unions, NUR/RMT and ASLEF. Between 1979 and 1999, it fell from almost 200,000 to 60,000[14]. In the old situation, the role of private business was very limited and marginal; everything was run by a public service. British Rail even did most of the maintenance and construction work. In the new institutional context, the role of the government–although not marginal–was very limited. Ultimately, the government made important choices about the definition of train paths.

The new institutional context achieved most of its aims. Fares went up only slightly, some by 2 percent, and the system's performance improved.

Since the first train networks were privatised in 1996, passenger volume has increased by 30 per cent, freight volume is up 48 per cent and 21 per cent more train miles are being operated[15].

However, system failures continue. Accidents and interruptions are frequent, and it is difficult to establish whether they are due to privatization and liberalization. Those interviewed said the interruptions are not worse than in the days of British Rail. However, the Cullen report following the crash at Paddington Station contradicts this, concluding that the industry lacked a clear leader in the area of the safety because of the disintegration. Public opinion attributed the interruptions, and especially the accidents, to the privatization and liberalization of the industry. This social resistance put a strong pressure on the rail revolution.

In addition to these failures, the industry proved unable to safeguard an important value of the new institutional context. The revolution was meant to stop the continuous flow of money to British Rail. However, Railtrack proved unable to keep its head above water. According to two respondents, *the incentives between the operators and Railtrack are misaligned*[16]. To safeguard its survival, the stock-exchange-listed organization received frequent financial injections from the government, which undid an important value of the new institutional context: clearly defined public and private roles. Eventually, Railtrack was taken off the stock exchange, its assets to be sold and its tasks to be handed over to Network Rail in October 2002. Railtrack was a private enterprise quoted on the stock exchange. Network

Rail is a not-for-dividend organization, commissioned by the government, and therefore an autonomous task organization.

3.4 THE ROLE OF KNOWLEDGE

The rail industry has an intrinsic need for technical coordination, first between vehicles and infrastructure, but also between vehicles themselves. This need is stronger than for cars, for example, because the technical role of the infrastructure in the rail industry is far greater. Infrastructure not only supports the vehicles, but also guides and steers them. History only increased the role of the infrastructure. The infrastructure started to supply electricity through overhead wires, set the speed through the security system, etc. This increased importance of the infrastructure led to ever-increasing technical coordination between vehicles and infrastructure.

Especially when innovations (e.g. in infrastructure, vehicles or services) have to be implemented, the technical coupling of systems prompts a demand for knowledge. The detailed knowledge of the subsystems has to be shared. Furthermore, this knowledge has to be applied from a shared responsibility for coordinating the subsystems, shared knowledge and responsibility going hand in hand.

The government often played a major role in aligning the fragmented interests of private enterprises round this technical coordination. Such coordination may be shaped as standardization. As early as 1842, the British government enforced the standardization of the railway gauge, enabling several vehicles to run on several tracks. Currently, the European authorities are standardizing traffic control systems. However, the development of the industry (higher speeds, greater reliability, greater flexibility and greater capacity), constantly requires extending and innovating the standards. Existing standards come under pressure when a new type of vehicle is introduced or a new infrastructure is built.

Privatization adversely affected both the level and the use of knowledge in the industry. First, privatization caused loss of knowledge of and skills in implementing rail transport. Privatization caused the various organizations in the rail industry to become leaner. The Strategic Rail Authority (2002) admits that "*experience and skills have been lost as companies have downsized*".

Second, knowledge in the industry is more fragmented than during the British Rail period. Fragmentation within the industry manifested itself in a large-scale procurement of rolling stock. Here, many old carriages had to be replaced as they were considered unsafe. These carriages, known as Mark I, had doors to every seat, which did not close and lock automatically. A large order was placed for rolling stock to replace it. The electrical systems in the new vehicles demanded more electricity than the old vehicles, which prevented them being used in large sections of the network by the time they were delivered. Coordination between the new vehicles and the infrastructure–particularly the electricity supply–was found to be lacking. It involved a great many parties: the vehicle producer, the ROSCOs, the carriers, Railtrack and OPRAF, the franchisor.

Third, enhanced risk awareness narrowed the tasks of the parties (see section 3.6), jeopardizing shared knowledge and responsibilities and further limiting the possibilities for successful innovations that should have made the revolution a success. Within the room managed by the parties themselves, internal innovations were realized, benefiting passenger numbers, for example. External innovations–those between parties–were slow to materialize, however, hampered as they were by fragmentation.

The system innovation transformed a single organization with many divisions into a large number of independent organizations. Although this did not create the problem of fragmentation, it formalized it. Knowledge sharing within an organization like British Rail is far from self-evident, and here, too, fragmentation of knowledge may be a problem. The planned and fixed nature of the system innovation strongly formalized the new relations, limiting the room for solutions in situations requiring knowledge sharing and knowledge development.

3.5 SOCIAL RESISTANCE

Many passengers have found that the system innovation was unable to put a stop to system failures. After the privatization and liberalization, they still face delays, trains are cancelled and carriers are incapable of a flexible response to, for example, the growing demand. The failure culminated in the two crashes at Paddington and Hatfield.

Resistance to the system innovation was growing, particularly to the privatization of Railtrack[17]. Passengers see the malfunctioning of privatized Railtrack as the main cause. Survivors underlined this in the media immediately after the crashes. They see the private nature of Railtrack as such as the cause. In a survey by the Guardian in October 1999, 70 per cent of those interviewed felt that Railtrack should be returned to public ownership. Other parties, too, are calling for the renationalization of the rail network[18].

The statistics do not directly warrant the growing resistance. They do not show any clear rise in the number of deaths by train crashes after privatization. Other performance indicators seem to indicate improvement: the quality of the service has improved, more passengers are carried and fares are not rising sharply[19].

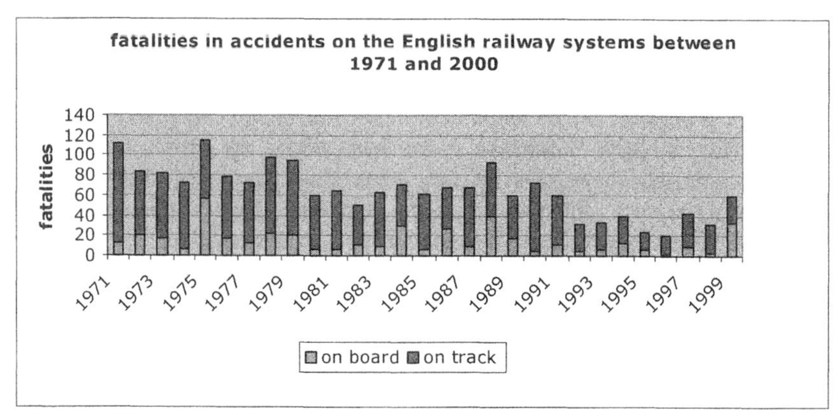

Figure 1 Fatalities in accidents on the English railway system between 1971 and 2000[20]

Whereas failure before privatization and liberalization was attributed to the monopoly of British Rail, failure after it is attributed to the removal of the monopoly by privatization and liberalization. In October 2002, this eventually led to Railtrack being restructured into Network Rail, a not-for-dividend organization taking over Railtrack's assets and tasks.

In addition, the social pressure led to growing attention being paid to safety. This is becoming the industry's "knock-out value", which society wants the government to embed in the industry. This has directly affected the franchisees. Stricter safety measures raise the operating costs of the franchises. The standards for vehicles, staff, infrastructure and internal management are rapidly becoming stricter. The carriers, although initially interested in developing the franchises over longer periods, are losing interest because of rising costs and risks. They now concentrate on performing their own limited tasks.

Doubtless, trade-offs between different values are made differently in a new institutional context. In the old context, professional incentives to improve quality, counterbalanced by limited financial incentives, dominated. In the new context, the tables seem to have been turned. Financial incentives now play a key role in the industry, counterbalanced by limited professional incentives. Carriers are held liable for accidents, and the strong role of financial incentives is used to create a new balance between professional and financial values.

The rearrangement of values in the industry explains why there is, or was, so much pressure on reversing or dramatically adjusting the system innovation. The aim of the system innovation was to improve the quality of service. Although this may seem to have been successful to some extent in, this advantage of comfort and quality is relatively invisible, even though a large group benefits from it. The failure of the industry, however, in interruptions and in accidents, is extremely visible, although it affects a relatively small group. As regards the cause of this failure, the finger is pointed at the institutional structure, and the system innovation is blamed for it.

Apparently, the system innovation has created its own resistance. The new institutional context has been politicized and associated with the Conservative end of the political spectrum from the beginning. The alternative form was indissolubly connected with the Labour end of the political spectrum. This old form of organizing the industry has never disappeared from view, offering spectators a continuous explanation for the failure of the system innovation.

The system innovation managed to address a number of visible shortcomings in the old institutional context. The innovation proved unable, however, to guarantee other values sufficiently, or at any rate make clear to users and the public that these values were guaranteed sufficiently.

3.6 PPS AND PARTNERING

Immediately after the privatization, the primary aim of partnering in the industry was to keep the costs of the infrastructure out of the government budget. Accordingly, builders and financiers in particular were involved in the realization of infrastructure. This partnering has the pattern of the Private Finance Initiatives: the government hires the facilities of "Special Purpose Vehicles" (SPVs), which finance, deliver and manage these facilities.

After 2000, the SRA tried to give carriers and other parties directly involved more influence in the development of infrastructure. The SPVs that the government then tried to set up were meant to be highly integrated. This form of partnering should be able to overcome the fragmentation when new projects were implemented. This broader form of SPVs evolved in an innovation project on the Chilton line and the construction of a number of underground lines. These broad SPVs were vertically integrated; carriers, builders, financiers and the infrastructure manager bore a risk.

However, this course was abandoned as early as 2001, because carriers showed hardly any interest. SPVs were again restructured. They were given a form strongly resembling traditional tendering, with the SPV regulating the relation between Network Rail and the SRA. The focus was again on funding: the new SPVs still made equity-financing possible, whereas, traditionally, debt financing was used when work was tendered out [21]. The advantages of equity financing were retained, but the government played an increasingly prominent role in the development of the infrastructure.

The planned system innovation led to a new trade-off between value systems. Planned system innovations like privatization and franchising of the railways risk focusing on some value systems in their design. The role of a PPS may be to counter this one-sidedness and couple different value systems to safeguard a value system that has become of secondary importance. Partnering gives other parties–and the values they propagate in the industry–a key role. Partnering may have a dual function. On the one hand, it allows a large set of value systems to be taken into account where choices demand it. On the other hand, partnering creates a wider variety in the way failure can be interpreted.

In the case study, however, the partnering initiatives failed to contribute an alternative approach to the industry. The financially-oriented partnerships also strengthened private values in the industry, hardly catering for professional and collective values. Those partnerships that did focus on them foundered through lack of interest.

3.7 CONCLUSION

The combined privatization and liberalization of the British rail industry may be regarded as a clearly planned system innovation. The industry was rearranged on the basis of a pre-developed desired final situation. In a two-party system like the British, solutions tend to be coupled to one of the two poles of the political spectrum. A political landslide tends to imply a revolution in the solutions chosen, in this case the institutional structure of the rail industry. It means that the existing relations are abandoned and new relations are introduced rapidly. The industry was changed dramatically in just over four years' time. The change placed most of the industry outside the public sector, limiting the new Labour government's possibilities for a "counterrevolution", also because the Blair government was more moderate in opting for drastic revolutions.

A major weakness of that desired final situation was the fragmentation of parties. The informal relations that used to exist within the British Rail organization were replaced by formal relations between the actors: the infrastructure provider, several vehicle providers, several carriers and several contractors and subcontractors for the construction and the maintenance of the infrastructure. The formal relations were part of the designed final situation. The contracts between the actors explicitly formed part of the desired final situation. They were drawn up to provide and pass on the right incentives to make rail transport more effective.

However, despite the extensive incentive structures, rail transport continued to face frequent failure. The new institutional structure proved to be little better or worse as regards failure. If there was a failure, there turned out to be no great difference in performance between the informal structures within British Rail and the formal structures after the system innovation. Other aspects of rail transport seemed to have improved, however, and fare increases turned out to be not higher than before the system innovation.

PPS-like institutional arrangements had already been present in the industry since privatization, but their aim was to generate private capital. They were arrangements that fitted within the Private Finance Initiative (PFI): private financiers create facilities and the government hires them back from the investors and developers for a fixed number of years. Such arrangements are called Special Purpose Vehicles (SPVs) in the rail industry.

Round the year 2000, the nature of the SPVs employed changed. Until then, they had been consortiums of builders and financiers, which we will refer to as the private SPVs in the rest of this chapter. In 2000, the SRA made an effort to involve carriers, too, in the SPV, thus expanding the number of parties. It was believed that involving carriers and others would enable better integration of knowledge between the parties to a project. Fragmentation by strong formal relations between the parties was regarded as a problem. A broad SPV for particular innovation projects would be able to cut through this fragmentation. Several of such SPVs were set up for the London underground. A single SPV of this type for mainline traffic was set up round the innovation of the Chilton line.

The system innovation introduced into the industry a formalization of relations between parties, increasing fragmentation between them. This seems to be a threat to an industry faced with major failures when some of the many couplings between systems and parties malfunction.

The SRA tried to mitigate the weakness of the rail-system innovation by means of broad SPVs, which would have to ensure the couplings between the parties round innovation projects. Two major crashes in 1999 and 2000, however, dashed expectations for these broad SPVs, causing higher safety standards for carriers, contractors and the infrastructure provider. This increased the cost of running rail services and even more that of implementing innovations. Carriers lost interest in the risky SPVs and again concentrated on providing transport services.

The solution for the failure, the broad SPV, itself became the victim of failure. This was due in particular to social pressure against the system innovation and in favour of the old institutional structure, even though the latter did not perform any better. Undoubtedly, the direct involvement of passengers and the visibility of the system innovation had an impact here.

The fragmentation introduced by the system innovation might have developed in two directions. First, the industry might have developed creative interactions where couplings in the industry demanded it. The alternative is an industry where cost awareness narrows the interpretation of tasks. The detailed design of the system innovation seems to have left little room for the industry to steer in another than the second direction. This was enhanced by the focus on safety that found its way into the industry after 2000, when a limitation of tasks clarified the responsibility and made it manageable.

The present British government does not opt for a new system innovation. It rather tries to find institutional changes that will remedy the main weaknesses of the present structure. The industry, however, is not always able to absorb possible improvements like the broad SPVs.

---

1. Beeching, 1963
2. Gibb et al.1996
3. Monami, E. 2000
4. A train path indicates the rail sections a train can use at certain times. The joint sections in place and time determine the speed of a train.
5. http://www.guardian.co.uk/traincrash/0,2759,180785,00.html

6. Lord Cullen, 2001
7. Monami, E. 2000
8. Interview 7 October 2002, Graham Eccles, Stagecoach Rail
9. Shaw, J., 2001
10. Interview 7 October 2002, Graham Eccles, Stagecoach Rail
11. Interview 8 October 2002, David Thomas, Strategic Rail Authority
12. http://www.guardian.co.uk/theissues/article/0,6512,744995,00.html
13. Perrow, 1999
14. Kay, 2001
15. http://www.independent.co.uk/story.jsp?story=332769
16. Interview 9 October 2002 Anson Jack and John Smith, Network Rail
17. Thompson, L.S. 2003
18. http://www.independent.co.uk/story.jsp?story=332769
19. Preston and Root, 1999
20. www.parliament.the-stationery-office.co.uk/pa/cm200001/cmhansrd/vo010123/text/10123w07.htm
21. Interview 9 October 2002. Derek Holt. OXERA

4

The Boston Biobang: An innovative system of biotech companies

4.1 INTRODUCTION

The developments surrounding modern biotechnology are often mentioned as an example of system innovation. Remarkably, no one explicitly manages this development. This is an emergent innovation, resulting from a chain of actions and reactions in the private sector. Although governments are monitoring the process, they did not initiate it, as in the other case studies.

The developments take place rapidly and are far-reaching for the respective market sector, regulations and underlying values. (More about this in section 4.3) This knowledge-driven innovation process is taking place worldwide and is too comprehensive to define in the context of this study. We will confine ourselves first to its application on drugs. Furthermore, we will focus on the driving force behind this process: the rise of a new knowledge and capital infrastructure on a regional level. For this purpose, we have spoken with a number of key actors in Boston and Cambridge, Massachusetts, a major biotech hub in the Unites States. The emergence of such regional infrastructures is a system innovation in itself. A large concentration of companies, knowledge institutes and venture capitalists has sprung up in a short period. Where relevant, we will also discuss the phenomenon on the federal level. In addition to Boston, other hubs like Stanford and San Diego have seen similar processes.[1]

Before discussing this regional level, we will give a brief introduction on modern biotechnology and its consequences for the pharmaceutical industry in the United States.

DNA, genes, proteins

The hereditary information of every organism is encoded in a long molecule: DNA. It consists of a huge sequence of bases, symbolized by the letters A, C, T and G. DNA is often represented as an almost infinite sequence of such letters, comparable to digital codes, all harbouring vital functions. It has been possible to explain the workings of some 'pieces' of DNA. We call such a 'piece' of code a gene.

On the basis of the information of the genes, the body produces *proteins*, which have the encoded functions performed. A deviation in a gene may lead to dysfunctional proteins, which are likely to cause a disease.[2]

Breakthrough

In the nineteen seventies, a breakthrough occurred in molecular biology. From then on, it was possible to distinguish, isolate and use genes. The dominant application of this discovery is the development of drugs.

Before this breakthrough, medicines were, and–until now–have been, developed to ‘repair’ dysfunctional proteins, without scientists exactly knowing the cause of this dysfunction. Pharmaceuticals develop and have access to an enormous amount of potential medicines. Specific diseases were screened against these potential medicines. So people looked for solutions for diseases and tried to find out what actually works. Drawbacks of this method are inefficiency, unpredictability of success and a big chance to miss the cause of the problem.

This process can be changed dramatically because of the scientific breakthrough. It is possible to know the cause of the ill- functioning proteins. Not medicines, but genes are the focal point of medicine development. Scientific research on the working of genes could lead to medicines against specific diseases or even a family of diseases.

This technological breakthrough has a number of conspicuous consequences:

1) New companies attract 'venture capital'
Every company managing a concept can thus potentially develop 'blockbuster drugs': drugs against a large group of diseases or against well-known diseases. If it manages to do so, that company may see exponential growth. This opportunity attracts venture capital, because this 'slim chance of great luck' appeals to some investors.

2) The type of patents is changing
Traditionally, patents are applied for on specific potential drugs. In many cases, it is still unclear by that time whether the respective drug is active against specific diseases and whether it will be marketed. The new technology makes it more attractive to apply for patents on 'discovered' genomes, because these are the starting point for drug development. The road from such a 'discovery' to the production of a drug (i.e. for approval) is long and complex. Such a patent may therefore have drastic consequences for every one involved in the development process.[3] In many cases, these are several companies and institutes. Currently, the question is being voiced whether such a type of patent is socially desirable. Meanwhile, a young company can subsist on a limited number of patents, because the 'venture capitalists' do not value them by their intrinsic value, but by their potential as a supplier of 'blockbuster drugs'. That potential is relatively large, for the same reason as the one mentioned above: the road to production is complex and may generate several active drugs.

3) Researchers change
Fundamental research used to be done by universities and applied research by the pharmaceutical industry. Now, university staff do fundamental research, and they themselves apply and develop it. When investors are sufficiently confident that a potentially practicable drug may be developed, they put up with temporary losses. 'Venture capitalists' (i.e. private funders) invest in young biotech companies, even if they incur heavy losses yet. Such companies tend to evolve round former university scientists and 'venture capitalists'. Their research may be characterized 'applied fundamental'. The pharmaceutical industry contracts out more and more of such research and tries to buy interesting companies.

4) The balance of power within the chain is changing
The chain from research to sale used to be dominated by the pharmaceutical industry. This was a limited number of medium-sized to very big companies. Several governments, such as states, the federal Food and Drug Administration (FDA) and the National Institutes of Health (NIHs), have interests throughout the chain. The states mainly stimulate innovation by offering favourable licensing requirements for innovative companies. The FDA indicates what safety standards have to be met before a product may be put on the market and what tests the drugs should pass. The NIH fund research projects and also buy drugs.

Developments after the breakthrough have led to an explosion of activity. A large number of small companies were established, funded by 'venture capitalists'. The big pharmaceutical companies face 'holes in the pipeline': the number of new products is declining.[4] This increases their dependence on the young companies' new products. Apart from the usual acquisitions, partnering is on the increase. Finally, the NIH and the FDA have difficulty in keeping up with the developments. What type of patents is acceptable? Should the testing procedures be adapted to the new situation? How should they deal with public resistance to genetic modification?

4.2 A BRIEF CHRONOLOGY

Having described a number of dominant, large-scale changes, we will now discuss a process that we regard as the driving force behind this system innovation: the rise of a knowledge and capital infrastructure on a regional level. We can describe this process in three phases.

Phase 1: A culture of poverty

The first biotech companies were organized round scientists. They tried to finance their activities with private money, doing so 'by all means'. Although they had boundless confidence in their invention and their mission to heal people, and although they invested a great deal of energy in it, it was difficult to attract funders. Far too little was known about the technology. Most of the money came from big, global companies, riskily investing a few per cent of their money. The universities themselves, first Harvard, then the MIT, gave facility support. The number of companies was small. Expertise in setting up facilities was both scarce and sporadic. This problem was met by cooperation and interaction.[5] In addition, the I-won't-take-no-for-an-answer attitude proved decisive in obtaining resources. In the early days, biotechnology was a world for tough types, for people suffering from the Marie Curie syndrome.

The first successes were scored in California. In Boston, two companies were eventually successful. Although they did not make any profit yet, they reached continuity in raising capital and put a number of drugs on the market.

Phase 2: First successes

These first successes were a flywheel for later companies. There was an explosive growth in the number of companies, increasing differences in quality. Companies benefited from the increased availability of capital and did not have to put products on the market until a much later stage. Even though they occasionally incurred substantial losses for the time being, the market value of some was extremely high, because they offered prospects for the future. Share prices of many of these companies were very high even while they were deeply in the red. However, the number of financiers specializing in biotech companies grew, especially when the first commercial successes were scored. Some pace setters, like Amgen, able to capitalize on a concept via one or more approved and successful drugs, served as an example to the rest, enhancing visibility and appeal. However, for investors lacking specialist knowledge it remained difficult to recognize the quality differences between different companies and concepts. Opening this black box was a job for superspecialists. Since not everybody was one, many uninformed investors suffered heavy losses.[6] Eventually, their favourite company failed to deliver the promised blockbuster drug, resulting in plummeting share prices, dragging their funders along.

Phase 3: Continuation

A clear turning point was the continuation of companies. They entered the limelight and appointed CEOs. They employed high-quality scientists, causing the pharmaceutical industry to accept them. Pharmaceutical companies increasingly started to partner with other companies contributing complementary knowledge and resources. They did so particularly when companies had drugs that were ready for testing but lacked the financial resources to perform the tests the FDA required before approving them.

The continuation impacted on the companies' strategic course. Originally, they were based on concepts and ideas. Now, they became more business-driven. Competition between companies increased, particularly when they focused on the same targets. The death rate among companies also rose. Although there were expertise and an infrastructure now, the focus was no longer on young enterprises. Given the vast number of companies, it was now difficult for an individual, small company to attract funders. In other words, the environment had become more familiar with biotechnology, and had greater confidence in it. However, there was now far more choice between young companies, which continued to compete for money. Expectations were more realistic and less high. The role models grew and bureaucratized. They received visits from legal advisors and increasingly began to resemble the big pharmaceutical companies ('Big Pharma') they used to rebel against.

4.3 THE TYPE OF SYSTEM INNOVATION

Values and institutions under pressure

The shifts described in section 4.2 were due to a chain of product and process innovations, because they concerned new drugs based on new concepts and produced by means of a new production process. These innovations caused great corporate activity, putting pressure on existing values and institutions, such as ethical values concerning technology. This technology assumed a new paradigm behind the treatment and manipulability of living organisms. The developments round biotechnology also affected values like public health and scientific values like public access to research findings.

Market conditions, which used to be stable for many years, changed because of the massive rise in small companies. Institutions attuned to the old conditions were in danger of losing value. Existing legislation about issues like public access to processes and patenting them were the constant subject of debate. Although pharmaceutical industry invested in the new companies, the latter would compete with them in the future. Governments tried to formulate and guarantee public interests. This was difficult, because much was unknown yet about the technology and its effects.

Drastic effects

Medical application of this technology had drastic social and institutional effects. First, the technology created opportunities for curing large groups of diseases, which gave many patients a direct interest in the rapid development of drugs.

However, the social effects were big, too: modern biotechnology not only impacted on public health. Other industrial sectors, like agriculture and the chemical industry, also used the new technological insights to their advantage. It is difficult to overestimate the magnitude of the effects in the future. Dramatic media reports about Frankenstein food, Dolly the sheep, the CopyCat and Raelian cloned babies frightened semi-attentive TV viewers and gave rise to an acrimonious social debate. It should be added that public support for the biomission is greater in North America than in Europe. North-American proponents of biotechnology like to point to food scams in Europe, which rather involved the 'old technology'.

System innovation emergent

This case study exemplifies an emergent system innovation in the private sector. Notably knowledge institutes and venture capitalists planned the developments. Although legislation played a crucial role, which we will enter into later, public institutes were stakeholders rather than designers of the process. We therefore suggest that the system innovation occurred on the interface between science and industry. Whereas the chapter about CERP had public parties acting as committed stakeholders and private parties as hesitant shareholders, the present case study deals with the reverse.

4.4 THE ROLE OF KNOWLEDGE

Scientific knowledge scattered in other disciplines

As the name suggests, the development of the whole range of activities round biotechnology–economically profitable in the long term–is essentially based on a combination of knowledge of microbiological processes and potential technical applications and interventions. Once scientists were able to map genomes and their components ('gene mapping'), modifying and applying them was no longer far away. This combination of insight into the elementary building blocks of life and the trial-and-error process of modifying them was of major scientific importance for a broad spectrum of fields. It not only affected biology itself, but also pharmacology, information technology, chemical process technology and agricultural technology, to mention just a few. Expectations are that currently only the tip of the iceberg is visible. Further development will make the effects felt in fields and disciplines where they are unsuspected at present.

The innovation did not spring from a growing consensus between those involved and authorities, as it did in the CERP case study, but from the application of a new insight or paradigmatic concept from microbiology to numerous areas of application. Relative consensus about genetics already existed, because it had its origin in biology, where advanced information technology was being used. This knowledge found acceptance elsewhere.

Divergence of knowledge from the supply

Because of the above factors, the system innovation in biotechnology is highly divergent. Talented scientists from renowned universities with high expectations and great ambitions are using their knowledge of and skill in genetics in areas as diverse as virology, drugs, embryology, crop improvement and chemical process technology, bringing about revolutionary changes in each of these fields. Hence, the initial source of the innovation is reasonably easy to define, but the distribution elsewhere and the consequences for each of the areas where this distribution takes place are impossible to predict. Remarkably, in the North American context, the innovators emanate unbridled energy: many of them categorically believe in their mission to improve the world by raising the quality of life and prolonging it by modifying the genetic material of other living beings and humans. In addition, reservations made as regards biological hazards of monocultures and ethical conflicts are relatively weak in North America, for reasons described under 'Resistance'. Although there are fears of adverse side effects of newly marketed drugs, the basic philosophy behind genetic modification of genetic information has long been embraced by the time that stage starts (see below).

Divergence of knowledge from the demand

The innovation was strongly research-driven. Caused by the great, fundamental, technological breakthrough, an infinite number of applications has been linked to it ever since. Arriving at a new concept or application requires considerable technological knowledge. Universities played a key role in the birth of the companies, and they still generate new companies.

A huge variety of new, potentially active drugs has evolved. Great progress is still being made in genome research. Much of it provides new concepts for drugs, and every concept has many potential applications. The large amount of potential targets leads to a constant demand for new applications. Divergence of knowledge is needed. The knowledge and capital infrastructure that have developed, as described above, allow rapid technological innovation, because money is available for good ideas, while the small-scale nature enhances speed.

Crucial knowledge on the interface of science and the market

Biotechnology has a number of important similarities with information technology as regards the rather divergent, evolutionary course of its development and its potentially revolutionary consequences. There is a large variation of ideas, solutions and applications, mutations take place rapidly and selection completely depends on usefulness on the market. However, apart from the ethical objections raised by some, there are two crucial differences:

- Developing applications is very expensive, requiring a great deal of money and excellent infrastructures (see above).

- Developing applications requires a very long time horizon, because approval by the public sector means that strict standards have to be met, involving a great deal of time, money and uncertainty.

Getting a new brilliant, or less brilliant, idea developed requires confidence and patience on the part of investors and funders. Investments are so high and the time it takes them to become profitable may be so long that a great deal of knowledge founders during the first stage. On the other hand, the commercial potentialities of some components of biotechnology are rated so high that many companies have high stock-exchange quotations without ever making a profit. This does not alter the fact that they will have to become profitable in the long term, or share prices will collapse. The question nevertheless remains how to separate the sheep from the goats? Experts either hardly know or do not know at all, and investors know even less. Books like the 'BioTech Investors Bible'[7] appear, describing the developments round the biotech companies from an investor's perspective. The interface between science and market is important for both sides, but it is full of uncertainty.

4.5 RESISTANCE

The long-term effects of modern biotechnology are both varied in form and largely unknown. Consequently, there is considerable resistance to it, especially on the part of consumer and environmental organizations, objecting to modern biotechnology with concepts like 'Frankenstein food'. Much of the resistance is religion-inspired, although it is remarkably mild in North America. The media have also paid a good deal of attention to recent publications for and against the new technology, much of it on philosophical or social grounds. One recent publication is that by Francis Fukuyama, who seriously objects to these excesses of the faith in progress and wants more extensive public debate[8]. Although some of their objections are justified and some professionals share them, biotechnologists have little respect for these opponents, because, they say, these critics do not understand. The resistance hardly hampered the rise of the new companies. There are a number of explanations for this, which we outline below.

Resistance to whom?

The first explanation has to do with the emergent nature of the system innovation. No party clearly initiated the innovation. The field of representatives of modern biotechnology is so fragmented that it is difficult to address resistance to one big, visible 'evildoer'. Resistance focuses on the federal government in particular. Safety is an important issue here. As a result, the FDA will hide behind its testing procedure, expensive for others, under pressure from federal politicians and public opinion. Relaxing this procedure in the interest of companies will be unacceptable to the FDA.

Tightening it tends to be the typical response to diffuse social resistance channelled through congressional representatives. This may require all manufacturers to dip into their pockets. It will also slow down the development without altering the basis in principle.

Incentives

As we said earlier, the innovation is taking place on the interface between industry and science. Representatives of both groups have an interest in new concepts and further development of existing concepts. Professional and financial incentives thus run parallel. This technology generates a great deal of new knowledge for those interested, and, potentially, a great deal of money.

In contrast with many other applications of modern biotechnology, the favourable effects of new drugs are both concrete and understandable: people are cured. The advantages are visible; disadvantages, if any, are invisible. This makes the medicinal application of modern biotechnology more acceptable to many.

It is easier for a smaller group of direct stakeholders to organize itself, and it has more convincing arguments available, because it has a direct financial interest and is well-informed. In many cases, the opponents work free of charge and are often wrong because of their scanty knowledge and expertise.

Knowledge-intensive

Because the subject is so knowledge-intensive, few action groups are expert in the area of modern biotechnology. It will not be easy for them to enter into debate with the proponents of the technology. A number of substantive mistakes in philosopher Francis Fukuyama's book have undermined his position: he often advances a one-to-one relation between a gene and a symptom, which, experts say, is simplistic nonsense. These weaknesses deprive his entire argument of its cogency, although it contains interesting general ideas about the social interests affected by modern biotechnology.

4.6 PPS AND PARTNERING

Public incentives in a private process

The system innovation is largely taking place outside the public sector. The knowledge and capital infrastructure evolved through market incentives of the most advanced kind: expectations about future profits based on 'blockbusters drugs', i.e. drugs against a major and well-known disease or drugs that can treat many diseases together. Venture capitalists may score immense profits, but have to exercise considerable patience. The same is true of most companies, some of which survive on their own, while many others are taken over by bigger companies. Although public organizations are not directly involved in the system innovation, they have major interests in the effects of the innovation. Some examples:

- *Safety*: The drugs that come on the market have to be safe. The FDA manages a standard procedure for testing potential drugs. It does so in three phases, with a rising scale and rising costs.
- *Compensation of market failure*: The market stimulates the development of 'block buster drugs', i.e. drugs with a high potential market value. There is a risk that applications for some less frequent diseases will fall out of favour, irrespective of the poignant consequences. The NIHs stimulate the development of such applications by means of subsidies. Like the FDA, the NIHs have 'technology transfer' programmes for this purpose: facilities are paid for and managed together.
- *Public access to knowledge*: Many new biotech concepts were developed by means of publicly funded research, i.e. by the NIHs. To prevent these concepts becoming company secrets, the Bayh Dole Act (1980) provides that findings from publicly funded research cannot be licensed exclusively.
- *Political programmes.* Some political programmes may match the biotech activities. The United States has the Ministry of Defense Biowarfare programme, which should ensure vaccinations for all Americans in case of bioterrorist action.

The public efforts referred to have a highly corrective level. Given the developments in the private sector, incentives are introduced to respect public values and/or procedures are enforced to guarantee them.

Effects on the innovation process: variety and partnering

Several instruments are employed to guarantee public interests, some of which have been mentioned above. Either intentionally or unintentionally, they influence the innovation process. The above-mentioned Bayh Dole Act, for example, has a substantial impact on the variety of companies: several companies can use a scientific discovery. The FDA procedure is also very influential. The procedure is too expensive for a young enterprise. It can therefore choose either to sell its concept in order to develop other concepts (horizontal integration) or to continue on the road to the market (vertical integration) and partner with a pharmaceutical company, because it is financially strong and has experience in FDA procedures. This has several consequences:

- The companies are dependent on the pharmaceutical companies. This dependence is changing, however. The production pipeline of the pharmaceutical companies is running dry because of the growing popularity of biotech drugs. This makes them more dependent on

partnerships with biotech companies, which can fill the pipeline.
- Medium-sized companies, too, cannot develop as many products as they want, because testing them all is too expensive. This limits the scale of the companies, until one company has succeeded in putting a critical mass of drugs on the market. Until then, they sell patents to other or new companies. This ensures great variety.

Private-private partnering for the most part, with participation of public stakeholders

In Boston and its surroundings, as in most other places in the world, the rise and further development of a complex of biotech companies and their environment is almost exclusively a private affair. Biotech companies themselves, investors, infotech companies, electronics giants and any Big Pharma companies interested in the commercial operation of specific applications enter into partnerships based on private aims that they try to realize together. With slight exaggeration, they might be called shareholders. Where public values are affected, public stakeholders will obviously make themselves heard. This rarely happens, however.

In private-private partnerships, we can distinguish three public or semi-public stakeholders.
- The most important of these, which, in a sense, created the Biobang in Boston and Harvard, are the *universities*. They supply knowledge and talented people. They often make facilities available, with the help of which applications can be developed. In a large number of cases, lecturers and professors also enter the market with their scientific ideas and try to develop them both technically and commercially. In other cases, they combine their position with running a biotech company. In a few cases, universities are partners together with the private parties. In such cases, they try to safeguard academic rights as well as copy and publication rights in the contracts. Usually, the other parties are after the patents, allowing an interesting division of the proceeds from cooperation.
- Second, *states* like Massachusetts and *municipalities* like Boston and Cambridge (Harvard) derive part of their attraction to industry and highly qualified and ambitious human potential from their existing presence. They have a high cultural level, good universities, promising employees and employers and try to recruit more of them. They are involved through lobbying and creating favourable conditions for business development. Municipalities like Worcester, situated in the same state but too far away from the innovation source, go out of their way to attract manufacturing companies in particular to get their piece of the pie.

- The third and final group of stakeholders is the *consumers* of products emerging from biotechnological development. Their interests (like safety and availability) are represented by several agencies, like the Food and Drug Administration (FDA) and the NIHs (see above).

Random examples of public-private partnership

Examples of public-private partnership do not exist in the Boston area, but they can be found on an international level. In many cases, they involve drugs, serums or viruses with a high social reward, but which are not, or cannot be, developed further from a commercial perspective. AIDS, highly prevalent in development countries, is one example of a disorder rated by commercial companies as too risky and/or not profitable enough. In some cases, governments or national and international social NGOs earmark financial resources to improve the chance of promising discoveries. In such partnerships, patents for countries in the developed world are allocated to the private parties, which put the product on the western market at high prices. The government or international societal organization receives the patent for the rest of the world, which is allowed to share the benefits of scientific progress free of charge, or almost so. As regards AIDS, this may be of great importance, particularly for Africa.

---

1. Howells (1999) speaks of 'regional systems of innovation'. We'll interpret 'system innovation' as the emergence of such systems.
2. For further reading: Dawkins, 1978 Ridley, 1999, Robbins-Roth, 2000
3. Jackson, 2003
4. Robbins-Roth, 2000, pp.165-166
5. Interview with Stephen Atkinson
6. Robbins-Roth, 2000, pp.111-118
7. Wolff, 2001
8. Fukuyama, 2002

5

A management view of system innovation

5.1 INTRODUCTION: THE SYSTEMIC NATURE OF INNOVATIONS

In this chapter, we will compare the three case studies. From this comparison, we will derive a number of insights about how system innovations develop and how we can manage them. We will also formulate lessons for a government that wants to realize a system innovation. We will do so in the form of observations and strategies. Observations may raise a government's awareness of system innovations; strategies offer possibilities for action.

When speaking of a system innovation, we can distinguish between the *substance* of the system innovation and the *process* of realizing it.

The classic picture of the *substance* of a system innovation is that it is both comprehensive and radical. The innovation is comprehensive, because the whole system should change rather than just a component of the system. This automatically makes the change radical: the system as a whole has to be revised rather than one component, or just a few components, while underlying institutions, values and norms also have to change.

A well-known example of this is the following. When an existing type of car is innovated to increase its ecological performance, we speak of an 'ordinary' innovation. Now suppose the innovation implies that the car will run on electricity and will be made suitable for automatic vehicle management. We then speak of a need for a *system innovation*, because such an innovation also requires adaptations to other components of the transport system, like petrol stations and the road infrastructure, which should allow automatic vehicle management. It also requires a change in norms and values. For example, autonomy is an important value for many motorists. They will have to sacrifice part of this value, because automatic vehicle management implies that motorists' autonomy will be restricted at certain times.

This leads to a highly rigid perspective of system innovations. Our research has led to a number of observations about the systemic nature of innovations completing this perspective.

When we pay attention not only to the substance of the system innovation, but also to the *process* shaping it, the term 'transition management' is common.[1] The term 'transition' denotes a period of some 25 years, during which system innovation takes place. Transition management can be realized in several ways, all of which appear to be based on the following principles[2]:

- System innovation is intentional: a 'transition goal' is formulated beforehand; this is a long-term aim;
- System innovation is deliberate: people are conscious of the systemic nature of the innovation and its implications.
- Interaction dominates: the transition goal is formulated together with relevant parties and constantly adjusted, because it is impossible for a single party or just a few parties to realize the system innovation.
- 'Backcasting' is used: backcasting is a technique that explores what concrete, short-term measures would fit the long-term transition goal. These measures in the short term will be guided by the transition goal.
- A prerequisite for system innovation is a change of paradigm. To bring this about, a major role is assigned to 'front runners', i.e. initiating parties with a lead in the direction of the transition goal.

Here, governments monitor the established substantive aim as well as the innovation process designed by other parties. This is referred to as a substantive role and a process role for governments.[3]

In this chapter, we will formulate a number of general lessons on the substance and the processes of system innovations. These lessons are based on the three case study descriptions. We will discuss:

- Realizing system innovations (section 5.2)
- The role of knowledge in realizing system innovations (section 5.3)
- Overcoming resistance in realizing system innovations (section 5.4)
- The role of public-private partnership in realizing system innovations (section 5.5).

We will then summarize these findings and draw lessons from them in section 5.6. There, we will also reflect on the principles of transition management.

5.2 SYSTEM INNOVATION: SOME GENERAL OBSERVATIONS

In this section, we will give a number of general characteristics of system innovations and classify them in broad terms. Every strategy for realizing a system innovation starts with the awareness of both general and special characteristics of a system innovation.

Observation 1: System innovations emerge on the interface of government, science or market

First, we can distinguish several types of system innovation. See figure 2:

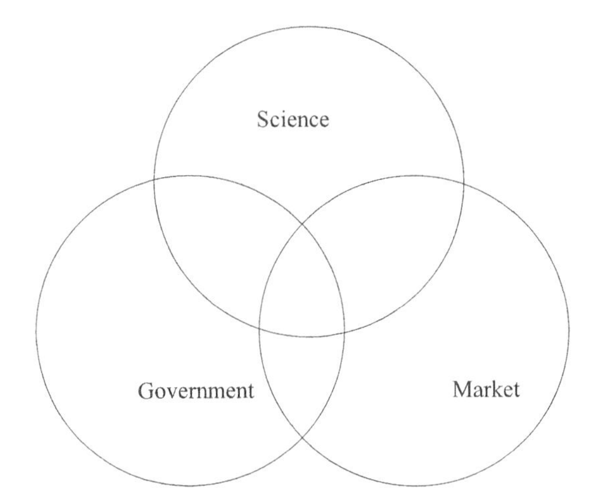

Figuur 2 Three sources of system innovation

- Knowledge-driven system innovations. This means that, from the arena of science and research, knowledge is generated about the need for and the possibilities of a system innovation. For example, research shows that a system is in danger of collapsing, and what measures are necessary to prevent this. From now on, we will speak of a 'scientific arena'. This includes science practised at universities and research institutes as well as research in companies and other institutes.
- Government-driven system innovations. The system innovation results from a political awareness that such an innovation is necessary and that there is sufficient support for and social acceptance of such an innovation.
- Market-driven system innovations. Possibilities for system innovations may also arise in the market; they tend to involve a coupling between new technological possibilities that are attractive from a commercial point of view. The IT revolution in the nineteen nineties is an example of a market-driven system innovation.

An interesting observation based on our case studies is that each system innovation described requires coupling between at least two of these three arenas. In the CERP case study, referred to below as 'Everglades', a coalition developed between the scientific and the political arenas, the Boston biobang resulted from a coalition between the scientific arena and the market, and the system innovation of British Rail took place on the interface between the political arena and the market.

In the literature, 'societal organizations' are often mentioned as sources of system innovation[4]. Our case studies have not found them a dominant source, which is why we will not include them in our analysis.

As we will set out in the rest of this chapter, it is highly relevant whether a system innovation is driven by government, the market or science and what type of coalition there is.

Observation 2: System innovations are not always clearly identifiable

Second, we have found that the systemic nature of an innovation is not always perceptible to the actors involved. We have found that the many incremental and spontaneous innovations that occur can only be characterized *ex post* as a system innovation. In one of the three case studies (viz. the Boston biobang), there is hardly any common aim guiding the system innovation. Although the privatization of British Rail was both deliberate and perceptible, most of the actual system innovation resulted from the dynamic that followed the decision to privatize it. The plan caused many unforeseen events, such as the failure of Railtrack, calling for new government measures. Subsequently, these in turn affected the new private parties, causing the original system innovation to keep changing. There was no ultimate perception of such significant implications of the design.

Observation 3: System innovations always have crucial emergent elements

In the third place, and as a consequence of the above: the system innovations have a large number of emergent elements. An emergent change differs from a planned change in having highly spontaneous elements. Within a system, seemingly automatic changes occur in certain components. They give rise to new changes, and eventually lead to the conclusion that the system as a whole has changed. Even in case studies in which a plan is of paramount importance in the realization of an innovation (e.g. Everglades), a large number of spontaneous actions and interactions precede the plan. At some time, they will result in a plan, a further incentive for the system innovation. However, this plan is not a closed plan: its implementation contains many unforeseen and emergent elements. There is no clear desired final situation, but there are strong incentives to keep changing.

Observation 4: The change strategy of the system innovation may vary: radical, organic or hybrid

Fourth, we found that a system innovation is not always as massive as some suggest and as was explained at the beginning of this section by means of the example of the electric car. A system consists of components (e.g. cars, petrol stations, the road infrastructure, the value of autonomy). From a massive perspective, a system can change only when all components change.

The British Rail case study clearly presents this picture: a comprehensive, radical and planned system change was introduced. A radical and comprehensive change was included in a design and implemented. It most resembles the intentional system change, as described at the beginning of this section, although, as we said earlier, it was followed by many non-intentional developments. All components of the system had to change. The opposite is the Boston biobang. Most of this system innovation evolved bottom up. It comprised activities like 'facilitating', 'bringing about' rather than 'introducing' and 'implementing'. This also resulted in a system change, involving changes in all components of the system, but it was far more emergent and organic.

Table 1 Three change strategies

	Strategy 1: British Rail	Strategy 2: Boston biobang	Strategy 3: Everglades
Who initiated the system innovation?	Government + market	Science + market	Science + government
What was the type of system innovation: design or emergent development?	Radical Comprehensive Unilateral design, followed by many emergent elements	Organic Emergent development	Hybrid model, both design and development
What was the central approach?	Top down: Starting from the desired final situation	Bottom up: All components changed as a result of bottom-up actions	Network: Starting from existing interests

Interestingly, the third strategy is far less massive than the first, although both were initiated by politicians. Had the first strategy been chosen in Florida, the intended change would have been designed as follows:

- There is the risk of drying out, which may affect the whole ecosystem of the Everglades.
- Consequently, a plan is made to prevent drying out.
- This plan requires changes to all other components of the system: different forms of using water, building, housing, agriculture, etc.
- Without these changes, the plan is doomed to failure.
- Instead, another strategy was chosen
- There is the risk of drying out, which may affect the whole ecosystem of the Everglades.

- We need to cater for other, legitimate values in this area: different forms of using water, building, housing, agriculture, etc.
- We have to find a new trade-off between these values, including the prevention of drying out.

Essentially, the difference is that strategy 1 starts from the desired final situation and the change it requires. This is a top-down approach: a government is able to define the desired final situation and impose it on other actors. The essence of strategy 3 is that the interests of the stakeholders are mapped first and that attempts are made to shape the system innovation so that it will conflict as little as possible with these interests. We call the third strategy a network approach, because these stakeholders form a network of mutual dependencies. Before adopting strategy 3, the Everglades used strategy 1. For many years, this provoked resistance and stagnation: the 'water wars'. This strategy is risky in a network. It demands ideal political and social conditions (see section 5.4 about resistance).

5.3 THE ROLE OF KNOWLEDGE

5.3.1 TWO DOMINANT PATTERNS OF KNOWLEDGE PRODUCTION

System innovations are knowledge-intensive. No system innovation will take place without knowledge of the existing system and the changes desired. In this section, we will set out what strategies of knowledge production we see. We distinguish two dominant patterns of knowledge production from the case study descriptions.

First, a pattern that we call *analysis/instruction* (section 5.3.2). This is the dominant pattern in the Everglades case study. Knowledge is generated from research and is then integrated. This leads to scientific findings about the desirability and possibility of system changes. Then, decisions are made, based on this knowledge. Proper *analysis* can *create* knowledge, which leads to more or less compelling *instructions* about the existing system, the need for a changeover to the new system.

We call the second pattern *variety and selection* (section 5.3.3). The production of knowledge shows market-like features here. There are a large number of knowledge producers, operating more or less independently. There is limited coordination of activities between them. Research is done on a large scale, which leads to a huge divergence of knowledge. Researchers also compete. In such an environment, there are constant incentives for knowledge production. Eventually, knowledge is selected from this variety and utilized.

In a market-like setting, knowledge production based on variety and selection dominates (e.g. the Boston biobang). This is a logical combination, because the market contains a variety of competing companies and researchers.

In a political environment, knowledge creation like analysis and instruction dominates (see the British Rail case study). This is another logical combination, because political decision-making should be based on the best and most complete information possible. Obviously, the strategy of analysis/instruction will dominate here.

In a science-dominated system change (e.g. the Everglades) both principles may apply, depending on whether the respective knowledge is felt to be of commercial and/or social importance. We will discuss the two patterns of knowledge creation in more detail in the following sections.

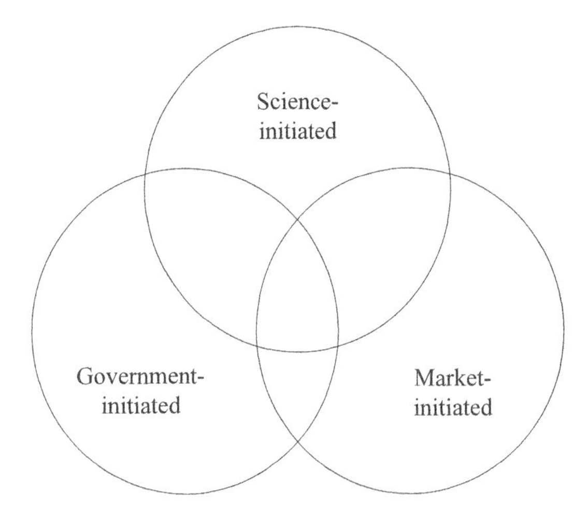

Figure 3 Three sources of system innovation

5.3.2 ANALYSIS/INSTRUCTION

How to create and diffuse authoritative knowledge?

We see the analysis/instruction pattern in the Everglades and the British Rail case studies. The strategy was the most prominent in the latter. This system innovation partly resulted from an intentional design, preceded by knowledge creation and analysis. The analysis was hardly contested because this was a top-down innovation. At some stage, the government was able to convince a sufficient number of other parties that the analyses of the existing and the new systems were right and that system innovation was therefore inevitable.

In the Everglades case study, the situation is different and less easy. The problem here is that the available knowledge is not integrated, that it is based on occasionally competing models and that it is too ambiguous to allow of a translation into instructions. In other words, the available knowledge is not authoritative and is therefore not supported by the main actors in the political arena. This problem is typical of a network-like setting, in which parties are interdependent and nobody can therefore impose the findings from an analysis on others. This makes a comparison between these two case studies interesting. In the Everglades case, an analysis/instruction pattern was followed, but how did the government manage to convince the parties of the correctness of the analysis and the inevitability of the instructions? What lessons can we draw from this? We have found a number of strategies that may answer these questions

Strategy 1: Framing and assembling: knowledge is framed from the stakeholders' perspectives

Knowledge is assembled and communicated so as to create sufficient new and attractive options for the existing stakeholders. They have their interests, and the available knowledge is framed so as to promote these interests. The knowledge presented also shows that the situation is serious, endangering the whole system. This creates a sense of urgency, the conviction that change is necessary and that it presents opportunities for the existing stakeholders.

Strategy 2: Negotiated knowledge: many investments in scientific consensus

Researchers invest a great deal of effort in consensus and the design of new, comprehensive models. Consensus and comprehensive models are *not* meant to underpin the need for and the possibility of change scientifically. They should boost authority in the political arena. This political function of scientific consensus is clear from the fact that

- such consensus is limited, and the models have a limited meaning;
- both the consensus and the models largely result from consultations between scientists and are therefore *negotiated.*

Strategy 3: Getting the attention of decision makers: simple, clear and 'sticky' message

Thanks to the consensus between scientists, their key message is relatively simple and clear: the ecological system is under threat; comprehensive solutions are possible that take all other functions of the Everglades into account. Such simplicity promotes attention being given to the knowledge in the political arena. It promotes the 'stickiness' of the message.

Strategy 4: Entice the decision-makers to act. The knowledge offers a clear perspective for action.

Finally, the knowledge is framed so as to offer a clear perspective for action. For all stakeholders, there is a prospect that serves their interests and contributes to the system innovation. Such framing removes the barriers to system change. Parties take action and are prepared to embark on the process of change.

Strategy 5: Creating knowledge is an ongoing process: the strategy leaves room for the next analysis/instruction round.

One major disadvantage of this approach is that reality is greatly reduced. Knowledge and the corresponding models are framed and reframed so as to be labelled as relevant to decision making. If there are any problems that the models cannot handle, they will present themselves at a later stage in the process of change. This is clear in the Everglades case study. Those that have to apply the models were not represented when they were developed. The problems they will face may threaten the consensus and therefore the progress of the system innovation. The above strategies will therefore have initiated a change process and broken an existing deadlock, but they will also pose a risk. Those who will have to apply the models in the future will need sufficient room to deviate from them. The system innovation may get bogged down if they lack this room. This implies that knowledge creation is an ongoing process and that new knowledge may affect existing insights. Although there is an analysis/instruction pattern here, it is a dynamic pattern. Analysis/instruction is followed by other analysis/instruction, which may differ from the preceding one.

5.3.3 VARIETY/SELECTION

Observation 1: A major change results from a large number of minor events

In the variety/selection pattern, there is a great variety of knowledge, from which the most relevant knowledge is selected. This selection process occurs frequently, but not necessarily, in a market-like environment.

An environment with a great deal of such variety is a breeding ground for system changes. At one stage, there will be so much variety as to enable a major breakthrough. In other words, a major change results in a large number of minor events. This process takes place as follows:

- First, the great variety of research allows a limited number of commercially interesting applications to be selected.
- If these applications are successful and have been selected as such, research looks promising and creates more variety. More resources become available, increasing the chances of another attractive selection.
- Finally, when there is a constant selection of applications, the sum of these applications will, one day, produce a system innovation.

For management, this means that the accent is on creating an environment or a context that leads to great variety and a good selection of knowledge rather than on creating knowledge. This is difficult, because such a context will have to match the dynamic of variation and selection. We will first discuss this dynamic, and then the potential strategies to make a selection from this dynamic.

Observation 2: The dynamics of creating variety: from an invention to a new knowledge infrastructure

An environment of variety largely develops because of a clustering of scientists in an environment that appeals to them. What are the determining factors? We will try to derive a number of patterns from the description of the Boston biobang. We see the following dynamic:

- The state of the art in a particular scientific discipline makes new breakthroughs possible. A sufficient number of scientists recognizes this. They intensify their research in the field in question, which creates a critical mass of research.
- Scientists attract scientists. Once scientists have set up in a particular place, they may create an environment that appeals to other scientists. This enhances variety as well as the chance of commercially interesting selections.
- At some stage, a government may contribute by facilitating this environment, by regulation, deregulation, physical infrastructure, marketing, etc., thus gradually shaping a knowledge infrastructure.
- A prerequisite for such a knowledge infrastructure is a potential capital infrastructure. Although the availability of resources is not an overriding factor, it may determine the speed at which variety is created.
- New roles emerge in this knowledge infrastructure. The scientists are the bearers of knowledge, the government facilitates and financial institutes or companies provide capital. A system change requires at least two other roles: *brokers* (i.e. organizations or persons that develop networks, for example between the researchers and the commercial companies) and *sales representatives* (i.e. persons or organizations able to commercialize findings, thus creating resources for new research).
- Finally, there is a factor x. Large-scale, scientific breakthroughs tend to give rise to the new culture: certain matters are cool, match the image of the new scientific activity and certain cities or areas offer that environment.

Observation 3: The dynamics of a selection process: selection creates variety

How does selection take place in the science-dominated environment of variety described above? We will try to outline the dynamic of selection processes in an environment on the interface between science and market, also on the basis of the Boston biobang case study.

- Some scientists were convinced that a new technology offered market possibilities. They will 'by all means' try to gain access to that potential market by further developing the technology on the one hand and interesting businesses in the technology on the other.
- Once companies have become interested, they start selecting, because they will select from the variety of research that which offers the best commercial opportunities. This is a self-strengthening mechanism: the

more selectors there are, the more incentives there will be to generate a variety of research..

- The selection environment will differentiate. A multitude of selectors develops, using different selection criteria. From a central perspective, the selection processes are therefore opaque.
- This differentiation results from different applications of the technology. However, the differentiation also generates incentives to retain or expand variety. This makes variation and selection an ongoing process.
- This ongoing process of variation and selection eventually leads to the system innovation.

Strategy 1: Coping with the dynamics of variation and selection: facilitating and coupling

The main implication of the above factors, determining the spatial clustering of scientists, is that they are extremely difficult to influence, not least for a government. We have managed to deduce three forms of influencing from the Boston biobang case study:

Regulating: regulations are adapted so as to create a greater variety of research. In many cases, innovative scientists face regulations that hamper risky research or even block it. A government may facilitate such research by changing these regulations, thus creating clarity and stability for scientists. For example, new regulations governing patents and licensing (the Bay Dohle Act) were essential for the research in Boston.

Facilitating: creating room. If a government notices a promising spatial clustering of scientists, it can facilitate the further development of that community. In such a case, facilitating means offering room for further development, enhancing the appeal of the conditions. American states try to create a favourable science and investment climate in their territories in the hope of sparking off variation and selection processes.

Coupling: going with the flow. In some cases, social priorities can be coupled with developments. The coupling between research in the biobang case study and the 'war on terrorism' (i.e. the bio-warfare programme) is an example. It is characteristic of this strategy that the 'managing actor' itself observes the rules that have evolved within the system. As it were, it participates in the innovation process as one of the many parties in order to achieve its own aims.

5.3.4 MANAGING THE ROLE OF KNOWLEDGE

We observed earlier that the analysis/instruction pattern occurs particularly in political decision-making, because political choices have to be accounted for in public, which is done best by means of an authoritative analysis, of course followed by instruction. The scientific arena creates knowledge, which provides instruction for the players in the political arena and eventually leads to actions. We recognize this pattern best in the British Rail case. Here, political decision-making is based on knowledge, both as regards substance and as regards process:
The knowledge warrants the decisions. Thorough analysis is important here: criticism of the analysis will undermine the authority of the political decision. This is why the knowledge has to be available early on in the process, enabling political decision-making.

In the Everglades, the system innovation evolved because of interaction between political decision-makers and scientists. Earlier, we called the Everglades strategy a network strategy. Both political decision-making and the analysis were attuned to existing interests. Knowledge was regarded as contestable, thus lacking the authority to serve as a basis for a political decision. In this section, we have mentioned a number of strategies for scientists and managers to arrive at decision-making within the analysis/instruction model while taking existing interests into account. Key phrases here are assembling knowledge, framing it and allowing it to take root.

In the Boston biobang case study, interaction between science and the market led to system innovation. We can describe this process best by using the variation and selection model. We described the dynamic of variation and selection in the light of this case study. Most of this dynamic takes place independently of governments. Variation and selection occur in constant interaction between science and the market. We then made a number of suggestions about how a government can match this dynamic. Key phrases here are: facilitating and coupling own interests to the dynamic described. Table 2 (right page) summarizes the above.

5.4 RESISTANCE

Because system innovations intervene, they may provoke considerable resistance, leading to delay, blockades or failure. Although this resistance may have a substantive background, considerations of power may be paramount. System innovations can easily harm existing interests, thus creating blockades. The question here is what strategies can address this. Such strategies may have been deliberately designed, but many of them have evolved more or less unintended.

Table 2 The role of knowledge in the three case studies.

	British Rail	Boston biobang	Everglades
How did knowledge for the system innovation come about?	Analysis/ Instruction	Variety/Selection	From Variety/ Selection to Analysis/ Instruction
How was this knowledge translated into action?	Emphasis on *creating* knowledge; this led to a *substantively good plan*, which was then *carried out*.	There was a variety of research, partly successful, *giving rise to* more variety. The most promising research was selected.	Emphasis on *assembling and presenting* knowledge, taking existing interests into account, thus creating a *plan* that was *attractive* to all parties. This was implemented
What was the essence of the strategy that had to be used?	Make sure there is a well-substantiated analysis.	Detect promising developments, couple interests, create a context (fiscal, regulatory) in which further knowledge production can flourish	Ensure consensus, 'stickiness', and good communication.
Where did the production of knowledge come in in the process of system innovation?	*Early* in the process, as a *basis* for the system innovation.	Constantly. Doing research and selecting was an *ongoing process* and eventually led to the system innovation	Early in the process, but the substance of the research and the way of presenting it was controlled by the interests of the stakeholders, who would come in later, in the decision-making

We compared the three case studies and found the following strategies.

Strategy 1: Widen the system boundaries

The first strategy is to widen the system boundaries. Although a system change is comprehensive already, in some cases it may be attractive to widen the system boundaries even further. The wider the system boundaries, the more components of the system are the subject of change. In other words, new issues are added in the decision-making process. See table 3. Such widening creates a "multi-issue game".

Such a multi-issue game is interesting because the new issues may lubricate the decision-making: First, they widen the playing field, allowing exchanges between actors. The more issues there are, the easier the process of wheeling and dealing becomes because the possibilities for exchange increase along with the issues.

Second, they form a strong incentive for cooperative behaviour, because multi-issue decision-making creates changing coalitions. When, for example, there are ten issues and five parties, the coalitions of proponents and opponents are likely to differ from one issue to another. This is an incentive for the parties to exercise restraint towards each other. Party A may be diagonally opposed to party B on issues 1 and 2, whereas they support each other over issues 3, 4 and 5. Because they need each other over issues 3, 4 and 5, they are likely to adopt a moderate stance over issues 1 and 2 rather than seek to win at any price.

This conclusion poses a paradox: the system innovation, which, by its very nature, is both comprehensive and radical, may benefit from expanding this complexity, because it creates a playing field on which changes can take place, including the system innovation.

In Florida, this strategy was called 'enlarging the pie'. Initially, the 'only' matter at issue was rehabilitating a nature reserve. These plans always conflicted with the water requirements of agriculture and the urban area. The plans were then reformulated, also taking into account the interests of farmers, Native Americans and urban water managers. This made these parties participate in the design of the plans and encouraged restraint.

Strategy 2: Find out the underlying professional and financial incentives for a system innovation and try to synchronize them.

Another important question is how the financial and professional incentives for the system innovation are divided. See table 3.

Table 3 Professional and financial incentives for system innovation

	Financial incentives negative	Financial incentives positive
Professional incentives negative		
Professional incentives positive		Favourable conditions for system innovation

On the one hand, there are financial incentives for or against the system innovation. When the system innovation offers major new commercial possibilities, these financial incentives will be positive, benefiting the system innovation. On the other hand, there are the professional incentives. Is the innovation interesting from a scientific point of view? Does it urge science to enhanced activity? If these incentives are also positive, a system innovation is likely. We can see this in the Boston biobang. The scientists were eager to innovate and, after the first successes, so were companies, because there were important commercial prospects. This synergy between professional and financial incentives creates an almost unstoppable dynamic.

Professional incentives tend to be a given, but particularly financial incentives can be influenced by third parties, including a government. One example here is the Everglades case study. In principle, the professional incentives were positive: the scientists agreed upon the direction of the system innovation. The financial incentives were negative, because it was an expensive project. However, for the Clinton administration, the environment was high on the political agenda, which allowed a substantial potential source of money to be tapped, which mitigated the negative financial incentives. This encouraged scientists and local governments to formulate a comprehensive plan quickly and get it accepted both by the federal government and by actors not sensitive to these financial incentives: the farmers, Native Americans and urban-water managers.

Strategy 3: Identify a dedicated target group. Make the advantages of the system innovation clear to this dedicated group.

In the third place, we mention the division of pros and cons of the system innovation among the actors. See table 4.[5]

Table 4 Division of pros and cons among actors

Proponents/ Opponents	Focused	Diffuse
Focused		Favourable conditions for system innovation
Diffuse		

The idea behind this figure is that a system innovation becomes more likely if the advantages benefit a dedicated group and the disadvantages hit a diffuse group. The Biobang case study clearly demonstrates this. The advantages of the innovation (e.g. curing diseases, prolonging life) involved a large group, but a small selection from it (viz. sufferers) had a paramount interest in it. The disadvantages of this development were far more diffuse: they involved potential future effects. Consequently, there tends to be a clear group of proponents and a diffuse group of opponents. It is an important mechanism that, in such a situation, the proponents tend to be more powerful than the opponents are.

We also see this in British Rail. The group against privatization, driven by accidents and disappointing punctuality, was clear. The interests of the proponents were diffuse. There was heavy resistance, resulting in the need to adjust the innovation.

In the Everglades, we found that the new insight (viz. water budgets can be doubled) let the benefits be divided equally among the direct stakeholders. The same was true of the disadvantages. It was doubtful, however, whether this new insight was tenable. If not, it would be unclear who would have to bear the cost of it. Although the benefits were thus postponed, the outcome was that there were no longer any major opponents of the plan.

Strategy 4: Make advantages of the system innovation visible

In the fourth place, we would point out the importance of the visibility of the intended system innovation. See table 5.

Table 5 Visibility of advantages and disadvantages

Advantages/ Disadvantages	Visible	Invisible
Visible		Favourable conditions for system innovation
Invisible		

The visibility of advantages and disadvantages has great influence on the resistance, even if the actual division of advantages and disadvantages is different. This, too, is illustrated by the Boston biobang case study. The advantages of new drugs were visible, the disadvantages were contested and might become manifest in the distant future.
Visibility was also found important in British Rail. The visible failure of the old system warranted innovation by the government. The new system, however, with its disappointing performance, suffered from the same visibility, which hampered the process of implementation . This case study also shows that perceptions are highly important. The causal links between the old or the new system and the failure of these systems are visible to many, but cannot be substantiated with facts. We should also point out a special mechanism in this case study. This innovation was a designed and radical system innovation, which provoked the tendency to blame it for anything that went wrong later, or was perceived to go wrong, such as safety. The failure of the system was highly visible and was linked to the system innovation, causing great resistance.

Strategy 5: Utilize momentum

System innovation is always a multi-issue game, because, many adjacent actors and factors may be found relevant. The strategy of widening the system boundaries adds to this. In such complex multi-issue games it is difficult to define beforehand when interventions will be effective. The aspects set out in this section may add momentum for system innovation.

We have observed earlier that the division of incentives may be subject to change. New insights may also influence perceptions of those involved and, consequently, the division of interests. Of course, it is important to remain alert to the above aspects. In British Rail, there seemed to be momentum: privatization was a discussion topic, there were complaints about the existing situation (where non-performance was very visible), the political constellation was favourable, (partly because of the two-party system), and there was an understandable design.

Table 6 presents the summary of the strategies we found.

Table 6 Resistance and the three case studies

	British Rail	Boston biobang	Everglades
System boundaries sharp or diffuse?	Sharp delineated system boundaries	Diffuse system boundaries	Diffuse system boundaries
What was the relation between professional and financial incentives?	System change aimed at overhauling existing incentive structure	Financial and professional incentives synchronous; system change used this incentive structure	Government boosted financial incentives, causing them to synchronize with professional incentives; system change used this incentive structure
How were advantages and disadvantages of the system innovation divided?	Disadvantages for focus group, causing great resistance.	Advantages for focus group, disadvantages diffuse.	Advantages for focus group, disadvantages postponed.
How visible were the advantages and disadvantages of the system innovation?	Potential advantages visible, like the 'manifest' disadvantages	Advantages visible, disadvantages invisible	Advantages and disadvantages moderately visible

5.5 PUBLIC PRIVATE PARTNERSHIP

What do the three case studies teach us about the role of public-private interaction in system innovations? There are two reasons why we can expect intensive public-private interaction in system innovations.

The first is the large scale of system innovations. Many of them will concern both private and public actors and require good and intensive public-private partnership.

In the second place, the literature about public-private partnership contains a number of key notions that warrant the expectation that system innovations will involve intensive public-private partnership.
Much of the literature advocates early involvement of private actors in public decision-making or vice versa, and a broad scope of the project.[6] This is believed to offer the best chance of fruitful cooperation and high-quality results. When public and private actors do not cooperate in a project until a later stage and the scope has already been determined, they tend to end up playing the traditional roles of, say, the government as principal and industry as agent. This leads to a separation of roles, whereas intensive bundling of and cooperation between parties is expected to produce better results.

Table 7 Common ideas about PPS

PPS from ...	To ...
Involving the other partner late	Involving the other partner early
Sharply delineated scope	Broad scope
Limited interaction; clearly separate roles	Intensive interaction

System innovation seems to match these notions, because the comprehensive nature of such an innovation means that the scope is broad and that intensive cooperation between public and private parties can be expected from the earliest possible moment.

Strategy 1: Don't involve partners too early: one sector in the lead, the other sector follows

A striking conclusion, prompted by all case studies, is that one sector is 'in the lead' during the initial phases of the system innovation: the public sector in Everglades and British Rail, the private sector in the Boston biobang. The other sector is subordinate. The explanation is as follows.

- *Private sector 'in the lead'.* The system innovation has its origins in the capillaries of market and science. Detecting this development and estimating its potentialities and magnitude require wide and profound

knowledge of scientific and market developments, which, in the initial phase, can only be found among those in these capillaries. Scientific breakthroughs like those round the Biobang occur at some stage and are partly unforeseen. For a government, such initial phases hold many uncertainties: in the initial phases, it lacks the in-depth expertise to appraise the breakthroughs and does not support them until a later time.
- *Public sector 'in the lead'.* A publicly initiated system innovation is comprehensive and involves many stakeholders. This creates a great many uncertainties: What direction will the innovation take? How will the decision-making go? How significant is a government pledge during the – often capricious – decision-making process? What are the consequences of a change of government? etc. Participating in such a process has little appeal for private parties. Here, too, there are too many uncertainties, either political or non-political, for companies to join, as this requires a minimum of consolidation and stability. This is the situation in the Everglades. In British Rail, we see a design approach, whose aim is to restructure the industry. In this design, the involvement of the private sector is limited.

Strategy 2: Develop intensive interaction, but after some consolidation of the system innovation

The above does not mean that the relation between leader and follower will remain unchanged during the whole period of the system innovation. Eventually, the sector that follows may play a facilitating role, thus boosting the system innovation. The Boston biobang is an example of this. The government facilitated this privately initiated system innovation at a later stage.

In this case, government had a limited, facilitating role:
- Either removing or erecting barriers by means of laws and regulations
- Creating fiscal, physical and spatial conditions
- Coupling the developments to other issues (Boston biobang: bio-warfare, drugs for the third world), thus raising public and political support.

The paradox is that this facilitating role requires very intensive cooperation between government and industry, because facilitating implies being on constant alert for new opportunities for, and threats to, the biobang, detecting these opportunities and threats and taking action to utilize the opportunities and ward off the threats. Figure 4 illustrates the development of the system innovation, contrasting an autonomous (i.e. private-sector) development (see the dotted line) with the innovation with public-private interaction. The dotted line represents the dominant development of the system innovation. It is the backbone of the innovation. Public interventions, as described above, may add something to this dominant market development.

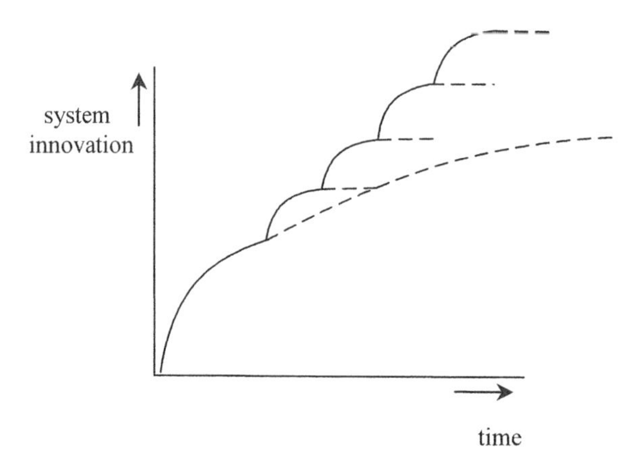

Figure 4 Public-private interaction and the Boston biobang

The publicly initiated system innovations (British Rail and Everglades) show a similar pattern. Initially, the governments were in the lead with a design for privatizing the railways and a comprehensive plan for the restoration of the Everglades. Once the decisions had been consolidated, the private parties moved in. The Everglades plan needed implementation, and new parties, some of them privatized, had to find their way in the rearranged rail industry.

This pattern is clearest in British Rail. After the system change, we see interaction evolving between government and private parties. Here, more than in the other case studies, a design was imposed on the industry. It turned out to be dysfunctional in several respects: investments in new infrastructural projects were too fragmented. For this reason, the government and private partners sought each other's assistance. The cooperation involved mending the dysfunctional consequences of the design. The figure shows the development of the system innovation. First, there is a sudden development because of the decision to privatize. The dotted line indicates the expectation as regards the course of the innovation process: once the plan has been implemented, the dust in the sector would settle, with new rules and new relations. Given the designed character, however, many adverse effects would present themselves after the decision. These required adaptations, for which the help of the private sector was called in. The system innovation was thus realized in strong public-private interaction, even long after the decision had been made.

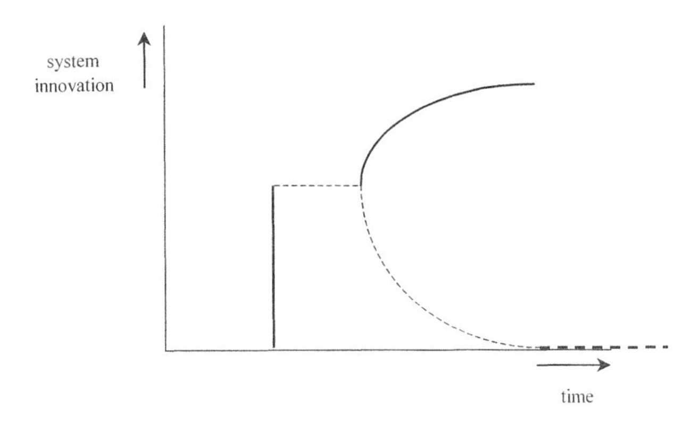

Figure 5 Public-private interaction and British Rail

The Everglades system innovation was a hybrid one. Although there was a comprehensive plan, it evolved in an organic interaction process. This makes the curve less abrupt than in the case of British Rail. 'Water wars' that had lasted for decades had to be overcome and autonomously developed, conflicting scientific models had to be coupled. The plan itself had to become attractive and understandable. Its comprehensive nature implied that it was full of assumptions and blank spots and that a host of unforeseen situations presented themselves. Later in the process, when the plan had been fixed, these assumptions and blank spots required intensive interaction between governments and, for example, private parties that had to implement parts of the plan and therefore had to adapt it. The great difference with British Rail is the realization of the comprehensive plan. Even before the political adoption of the plan, there were breakthroughs that made system innovation possible. The plan therefore evolved far more organically.

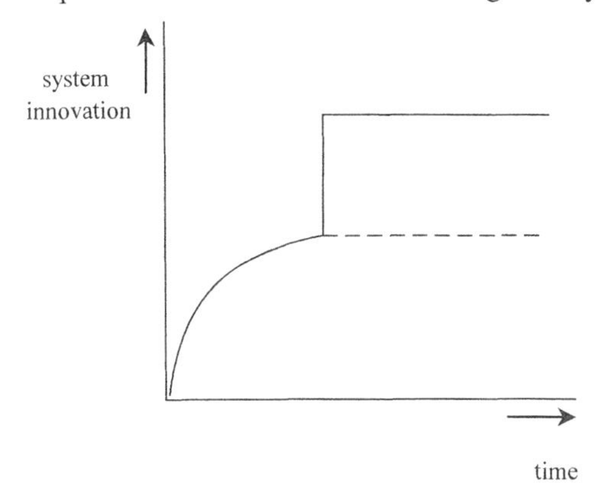

Figure 6 Public-private interaction and Everglades

Table 8 The three case studies and PPS

	British Rail	Boston biobang	Everglades
What sector was initially in the lead?	Public sector in the lead. Private sector initially lagged behind, because the design was fixed unilaterally; joined the process at a later stage.	Private sector in the lead. Public sector initially lagged behind because there was no in-depth expertise; joined the process at a later stage.	Public sector in the lead. Private sector initially lagged behind, because the political decision-making process was too capricious; joined the process at a later stage.
Was the scope fixed or in motion?	Scope fixed in plan, but in motion after implementation	Scope in motion	Scope in motion, coagulated in plan, in motion after being fixed
What was the focus of the public-private partnership?	Implementing and mending	Guaranteeing public values and facilitating further development	Implementing and adapting the plan

The patterns we see in the rise of PPS and the type of PPS are easy to explain when we remember that the Boston biobang was an organic system innovation, British Rail a designed system innovation, and the Everglades a hybrid one. Although there was a comprehensive plan, it resulted from an organic process of interaction and required extensive interaction in its implementation.

In an organic system innovation, in the capillaries of the market and science, the private sector is obviously in the lead. The public sector follows and will facilitate the organic development. PPS usually aims to safeguard public interests, like safety and public health in the biobang. These roles require intensive interaction between government and industry. The scope of the system innovation is in constant motion; new opportunities constantly present themselves.

In British Rail, where the system innovation was designed, the government was in the lead, establishing the design more or less unilaterally and implementing it. The original scope was laid down in this plan. Naturally, given the magnitude of the change, unforeseen problems arose. The solution of these problems required a repair-style public-private partnership and interaction between government and industry. The final scope resulted from this ongoing interaction and was therefore in motion.

If, in a hybrid form, the public sector is initially in the lead, companies are unlikely to join at an early stage because of the capriciousness of decision-making in system innovations. Private parties come in once the plan has been adopted. Because the plan is so comprehensive, it cannot just be implemented, but it requires constant adaptations. This, in its turn, leads to intensive interaction between public and private parties, with a constant shift in scope.

Strategy 3: PPS as the management of competing values

In an earlier publication, we identified eight competing values that may present themselves when public-private interactions are designed and managed.[7] We briefly introduce four of them here:

- *Designed or spontaneous?* Should a party design the PPS from scratch or should PPS relations evolve organically?
- *Demand-driven or supply-driven?* Does the public-private partnership spring from a demand in the market or are the public-private activities supply-driven?
- *Fundamental or applied?* If the PPS is based on research, is this research fundamental or is it application-driven?
- *Quality or relevance?* If research has been carried out, what decides the value of it? Its quality, or its relevance?

We call these tensions competing values, because both values in each spectrum may be worth pursuing: both contain positive elements. These tensions will therefore continue to surround PPSs. Formulating such competing values facilitates the understanding of PPSs and characterizing them.

We have found these competing values in the case studies. PPS is not a matter of opting for either of the values in each of these dimensions, but of constant toing and froing between these values, because one-sided orientation towards one of these values on a spectrum would disregard the fact that the competing value also has its positive aspects and should not be neglected.

We will explain the management of competing values by these four tensions. First the 'designed or spontaneous?' spectrum. PPS has strong, spontaneous elements. One of the sectors is the follower in the development of system innovation and gradually becomes involved in it. Once the public-private interaction phase starts, part of this interaction will institutionalize, and the PPS will also adopt design-like elements: deliberate and planned interaction.

On the 'demand-driven or supply-driven' spectrum, we also see that both values play a part in the PPS. Pressing social issues are at stake, and the PPS is therefore largely demand-driven. The PPS also creates new, unforeseen possibilities, which are supply-oriented. This is very clear in the biobang case study: it is neither purely supply-driven (the supply of solutions), nor purely demand-driven (the demand for drugs), but there is constant interaction. Something similar applies to the competing values 'fundamental or applied'. The research in the biotech case study has both fundamental and application-based elements; it is intermediate between fundamental and applied research, depending on the stage of the research. Fundamental research findings are applied, application-based research requires a fundamental basis to allow it to develop further.

Finally, also on the 'quality – relevance' spectrum, not one of the values is chosen, but there is a constant toing and froing. This is clear in the Everglades case study: the relevance of the research is such a dominant value that less attention is given to quality. Thanks to this choice, the research has great impact, but a great deal of attention has to be given to the quality of the research later. This is because, after the political decision-making, certain assumptions and models are found not to work in the implementation and have to be adapted. The choice of quality alone would have produced good results, but no impact. The choice of relevance alone would have resulted in great impact, but poor implementation. Here, too, there is a moving back and forth between these competing values.

When shaping PPS, it is important for the parties to be aware of the existence of these competing values and to realize that one-sided choices are dysfunctional. Failure by the parties to realize the need for this management of competing values may create considerable confusion and seriously disrupt the process of public-private interaction.

In line with the preceding observation, none of the three system innovations had a central PPS construction. PPS tends to be peripheral to system innovation. PPS usually aims to safeguard values that are in danger of getting snowed under by the new trade-off. This is clearly visible in the Biobang case study. Commercial values conflicted with the idea that everybody should have equal opportunities to obtain drugs. PPS evolved because of the costs of drugs in development countries and governments stimulated the development of drugs that were less interesting commercially.

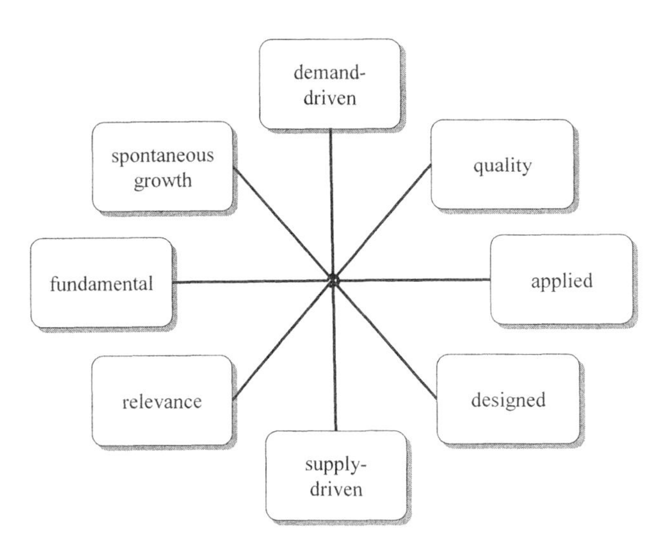

Figure 7 Competing values in designing public-private partnership

5.6 SUMMARY AND REFLECTION: CREATING SYSTEM INNOVATIONS?

On the basis of three case studies, we have distinguished a number of observations and strategies that may be useful for governments that seek to manage system innovations. In this chapter, we will summarize these case-study-related observations and strategies. We will then try to transcend the case studies by presenting a first ordering in potential tools. Table 10 summarizes the observations and strategies. The table follows the structure of the sections in this chapter:

- How did the system innovation come about?
- What was the role of knowledge in the system innovation?
- How was resistance dealt with?
- What was the role of public-private partnership in the system innovation?

Table 9 Three strategies for system innovation

	British Rail	Boston biobang	Everglades
Realizing system innovation			
Who initiated the system innovation?	Government + market	Science + market	Science + government

(table continues)

(table 9 continued)

What type of system innovation was it?	Radical Comprehensive Unilateral design, followed by many emergent elements	Organic Emergent development	Hybrid model, both design and development
What was the central approach?	Top down: Starting from the desired final situation	Bottom up: All components changed because of bottom-up actions.	Network: Starting from existing interests
The role of knowledge			
How did knowledge for the system innovation evolve?	Analysis/ Instruction	Variety/Selection	From V/S to A/I
How was this knowledge translated into action?	The focus was on *creating* knowledge; this led to a *substantively good plan.* This was *carried out.*	There was a variety of research, partly successful, *giving rise* to more variety. The most promising research was then selected.	The focus was on *assembling and presenting* knowledge, taking existing interests into account, thus creating a *plan that was attractive* to all parties. This was carried out.
What was the essence of the strategy used?	Make sure there is a well-substantiated analysis.	Detect promising developments, couple interests and create a context (e.g. fiscal, regulatory) in which further production of knowledge will thrive.	Ensure consensus, 'stickiness', good communication.
When did the production of knowledge feature in the process of system innovation?	*Early* in the process, as a *basis* for the system innovation.	Constantly. Doing research and selecting was an *ongoing process,* eventually leading to the system innovation.	Early in the process, but the substance of the research and the way of presenting it was controlled by the interests of the stakeholders, who would come in later in the decision-making.

(table 9 continued)

Dealing with resistance			
System boundaries sharp or diffuse?	Sharply delineated system boundaries	Diffuse system boundaries	Diffuse system boundaries
What was the relation between professional and financial incentives?	System change aiming to break existing incentive structure.	Financial and professional incentives synchronized; system change used this incentive structure.	Government strengthened financial incentives, synchronizing them with professional incentives; system change used this incentive structure.
How were the advantages and disadvantages of the system innovation divided?	Disadvantages for focus group, causing great resistance.	Advantages for focus group, disadvantages diffuse.	Advantages for focus group, disadvantages postponed.
How visible were the advantages and disadvantages of the system innovation?	Potential advantages visible, as were the 'manifest' disadvantages.	Advantages visible, disadvantages invisible.	Advantages and disadvantages moderately visible.
The role of public-private partnership			
What sector was initially in the lead?	Public sector in the lead. Private sector initially lagged behind, because the design was established unilaterally; joined the process at a later stage.	Private sector in the lead, Public sector initially lagged behind because it had no in-depth expertise; joined the process at a later stage.	Public sector in the lead. Private sector initially lagged behind, because political decision-making process too capricious; joined the process at a later stage.
Was the scope fixed or in motion?	Scope fixed in plan, but in motion after implementation.	Scope in motion.	Scope in motion, coagulated in plan, in motion after plan was adopted.
What was the aim of public-private partnership?	Implementing and mending.	Guaranteeing public values and facilitating further development.	Implementing the plan and adapting it.

These observations and strategies were identified from concrete case studies and therefore have a certain logic and consistence.

- The plan in the British Rail case study was designed by a government in a relatively unilateral way and introduced top-down. The knowledge pattern will therefore closely resemble analysis/instruction, because political choices have to be answered for in public, which is done best by means of an authoritative analysis, naturally followed by instruction. It is also natural that the plan will provoke great resistance after the decision. Finally, public-private interaction because of the decision will increase due to this resistance and imperfections in the design.
- The system innovation in the biotech industry was realized bottom-up in a market-like environment. The knowledge pattern will therefore resemble variety/selection, because selection was decentralized, made by a multitude of actors rather than a government on the basis of a multitude of criteria. Parallelism between financial and professional incentives is a condition for the emergence of such a process. Parties in the public sector will want to join the process by combating imperfections on the one hand and coupling their interests to the developments on the other hand.
- The Everglades case study is a hybrid form. We have called the scientists' strategy a 'network strategy'. It acts from existing interests. The pattern of knowledge production and dealing with resistance is adapted to this strategy. Knowledge production focuses on the link with the existing interests: how can we frame knowledge in such a way that each party can gain something? The same is true of handling resistance: the plan is framed in such a way that there is something in it for everybody. There is a great deal of interaction, which adopts a highly public-private character during the implementation of the plan.

Our main conclusion is that all system innovations contained many emergent elements. In the instances examined, governments played important but very different roles. The emergent development of system innovation would seem to conflict with the idea of a proactive government. In conclusion, we would like to set out what role a government can play if a system innovation is predominantly emergent. We will seek the answer to this question in two directions:

1. Develop tools and strategies that take the emergent nature of system innovation into account.
2. Make existing tools and strategies 'emergent proof'

1) Developing tools and strategies

From the case descriptions, we have identified a number of strategies that would seem applicable in a number of circumstances. Of course, the following list is not exhaustive, but it gives some impression of the type of tools that may be effective in managing system innovations.

Strategy 1: Create porous boundaries between science and the market.

In the Boston case study, we see porous boundaries between science and industry. A porous boundary means that there is ample interaction, cooperation and bundling: joint projects are carried out, scientists place social or corporate issues high on the research agenda, establish companies in which they commercialize their results, etc. Second, within science there are porous boundaries between units (e.g. between fundamental science and applied science, between different disciplines). The idea is that the chance of innovation is greatest on the boundaries between disciplines and/or between fundamental and applied science.

Porous boundaries are a major prerequisite for the emergence of system innovations between science and market. For example, in the United States, scientists abandoned their academic careers to set up a biotech enterprise. The large scale on which this happened gave rise to system innovation. Why did it take place in Europe on a much smaller scale? This may be because there are fewer professional and financial incentives for entrepreneurs: science in the Netherlands has a relatively high status, whereas in the United States it is easier to amass a fortune in business. Another relevant observation is that European universities are organized in "faculties" (i.e. schools), which hampers innovations on the interface of disciplines.[8]

Strategies to remove such barriers can substantially stimulate system innovations on the interface of science and the market.

Strategy 2: Facilitate an ongoing process of variety and selection.

Strategies may be aimed at creating individual variety or at influencing selection processes. The Boston biobang case study shows, however, that–under certain conditions–the selection environment may form an incentive for variety and that variety can influence the selection environment. This makes variety and selection an ongoing process. A government can facilitate it by creating conditions under which processes of variety and selection may thrive. It can do so on the one hand by creating stable frameworks by means of laws and regulations that offer clarity to parties from the market and science. It can also do so by facilitating these parties individually, offering a favourable climate, either financial or otherwise.

Strategy 3: Organize the process of detecting system innovation.

Links between science and the market will always be diffuse and difficult to design. The metaphor of the plate of spaghetti thrusts itself upon us: there are many links, closely interwoven. This takes us to another answer to the question about facilitation. Of course, a government may try to create incentives that promote the realization of this type of bundling, like–to give a few simple examples–funding interdisciplinary research or expanding the possibilities for scientists to commercialize research findings. This is a form of ex ante facilitation, aimed at creating a breeding ground for innovations.

Such forms of ex ante facilitation may supplement ex post facilitation, like subsidies for system innovations that are already in progress.

Ex ante facilitation demands a government that is alert to potential system innovations and, where necessary, creates conditions for the above-mentioned bundling without being asked. As we said earlier, not the substance of facilitating is paramount, but the process of identifying system innovations and worthwhile public facilitation. This also requires an enterprising attitude. Designing arrangements to facilitate innovation (e.g. funding schemes, tax credits) is easy enough, whereas identifying in good time where they are needed is difficult. Detecting them will not only be a task for one single government agency, but for those directly involved: network-like organizations on the interface between science, government and the market.

Strategy 4: *Make sure that professional and financial incentives run parallel.*

We have found that the incentive structure is essential for the speed of a system innovation. This finding suggests that the use of financial tools is effective when professional incentives are positive. This, too, requires a strategy of detection (strategy 3). It is very important to know where there are new (professional) opportunities and interests.

Strategy 5: *Freeze the process of system innovation where necessary...*

Many regard emergent processes as chaotic. There is no clear leader setting any direction. The outcome of the process is unclear. Thus, it is difficult to anticipate an outcome and to invest in it. An emergent process does not offer the parties enough certainties. For example, one of the respondents suggested that this is one reason why public-private partnership in such processes is slow to start.[9] We take the view that, in every process of system innovation, all parties will feel the need for some freezing: a conclusion on which to design further innovations.

Freezing may have the form of government regulations. Regulations accepted and complied with by all parties can be a substantial framework for the development of a system innovation. We frequently hear the call for regulations, such as those on genetically modified crops or major infrastructural interventions.

Strategy 6: *...but use process-based reasons when doing so.*

The need for government regulations can be both strong and diversified. Commitment to a rule, and, consequently, the freezing potential of the process, largely depends on the time when the rule becomes effective. If there is no commitment, there is a risk that regulations will provoke a dynamic like that in the British Rail study rather than freeze a process.

Of course, there are various forms of freezing, such as the development of technical norms, standards and agreements between suppliers of innovative products. Commitment is essential for all forms of freezing. It is for those involved to find the momentum at which decision-making can be consolidated. For science and the market, this consolidation will hardly be spontaneous. By definition, market parties are numerous and difficult to align. The research field is relatively diffuse and difficult to coordinate. Government regulation is vitally important:

- Governments can initiate freezing by means of regulation. They will have to find the momentum themselves and, if necessary, create commitment themselves (For this, see the notions about transition management).
- They can facilitate the fixing of standards between parties by, for example, simplifying regulations.
- They can initiate and facilitate agreements between parties, possibly under pressure of regulation. Examples are the realization of the partnership agreement on packaging[10] and the one on benchmarking energy efficiency.[11]

2) Making existing strategies 'emergent-proof'

Earlier, we described a number of dominant concepts of transition management. In these, system innovation is intentional and deliberate. The system innovations described saw an emergent development and were, at times, difficult to identify for those involved. These tensions also affect the practice of transition management, like formulating transition goals, the use of 'backcasting', and encouraging front-runners. The following elements from the case studies conflict with the concepts of transition management:

- 'Backcasting' will be difficult in the interaction between science and the market in the event of a system change, because in such a change there is no collectively formulated goal, but a composition of fragmented goals. If the field in which the change occurs is highly fragmented, as in the biobang study, such collectivity is difficult to realize.
- A plan initiated and laid down by a government may be an important component of system innovation, but such plans tend to trigger an unpredictable dynamic that should be regarded as part of the system innovation, as in the case of British Rail. System innovation always contains emergent elements, which may raise the same questions about 'backcasting'.
- System changes are difficult to identify. How systematic are the changes? What will they lead to? If it is impossible to find clear answers to these questions, it will be difficult in the first place to recognize the desirability of a change of paradigm and manage it with the help of front-runners. Second, it will be difficult to identify front-runners in advance in order to encourage them. An illustration of the latter from the Boston biobang case: with hindsight, it is clear that big, successful biotech companies like Amgen were front-runners, but they gained their

lead in an uncertain process, in which many similar companies threw in the towel. Initially, it was unclear what consequences this process would have.

The case studies are no reason for optimism about the management possibilities for governments. They question the value of the realizations of transition management for such emergent processes. We would suggest that this depends on the way these realizations are used. This is the challenging mission of a transition manager, who has to strike a balance between the parties' management ambitions on the one hand and the emergent development of system innovations on the other. Of course, striking a balance is not easy and depends on the conditions of the process. Table 11 sets out this dilemma.

Table 10 Two realizations of transition management

Realization 1	Realization 2
Function of the transition goal: controlling	Function of the transition goal: committing
Narrow transition goal	Broad transition goal
Backcasting: maintaining transition goal	Backcasting: learning about transition goal
First forecasting, then backcasting	Backcasting and forecasting in tandem

If management ambitions are high, actors may feel inclined to give the transition goal a controlling function. The goal should guarantee that management ambitions will be implemented. These have to be laid down in the transition goal beforehand. If the circumstances dictate moderation of the management ambitions, a process function of the transition goal remains: an aim may serve to continue to commit actors to the transition process and possibly commit new actors during that process.

The function becomes explicit in the formulation of the transition goal. A transition goal may have a narrow formulation, with the advantage of providing clarity and indicating a direction. A broadly formulated transition goal, however, is more flexible and appeals to more parties. A broad transition goal will therefore generate more commitment, although too broad a transition goal is meaningless and will commit nobody. An example of a narrowly formulated transition goal is ‘reducing meat consumption by 50%’. An example of a broader transition goal is ‘improving environmental efficiency by 50%’. The second aim leaves more room for the way the aim should be realized.

The function and ambitions of a transition goal have a major impact on the way ‘backcasting’ can be used. If a transition goal should guarantee the management ambitions, 'backcasting' may adopt the nature of enforcement. 'Backcasting' then becomes a guiding principle for those involved, which may make the memory of the transition goal a management incentive for those involved. On the other hand, 'backcasting' may also be given a more modest function, a learning function. The feedback to the process of system innovation is used to learn about the transition goal. This learning process not only concerns the substance, but also the view that those involved take of the present process and the transition goal.

The relation between 'backcasting' and its counterpart 'forecasting' is closely linked to the preceding. A transition goal is fixed by means of 'forecasting'. Subsequently, meaning is given to the current transition process by means of 'backcasting'. When actors adhere to the transition goal, 'forecasting' will hardly play any further role. However, when the status of the transition goal and 'backcasting' is more modest, the process will acquire meaning by a constant combination of 'forecasting' and 'backcasting'. In other words, the question what the present state of affairs means for a transition goal has the same status as the question what a transition goal means for the present process.

The more emergent elements a process has, the more modest the functions of the transition goals and 'backcasting' will become and the more transition management will be realized in line with the right column. It is doubtful, however, whether present management ambitions can be implemented, if transition management completely follows 'Realization 2'. For this problem, we refer to the strategies described earlier in this section.

Both the new strategies with their corresponding tools and the existing strategies are highly process-oriented. In so far as they have a substantive or a hierarchical connotation, the reasons for it are process-related . This is not so surprising, given the context in which the system innovation takes place. This is a context in which there is no authoritative knowledge, consensus is not a matter of course and tension exists between public and private values (see chapter 1). Processes of system innovation are therefore capricious and unpredictable, as our case studies show. Strategies that are either too substantive or hierarchical lack the ability to anticipate such emergent processes.[12] Undoubtedly, their substantive underpinning will be contested, they will show imperfections and, in their turn, cause unpredictable dynamic.

---

1. Rotmans et al, 2000
2. Netherlands Ministry of Economic Affairs, 2001, Sustainable Technological Development Knowledge Transfer and Embedment, 2000, Rotmans et al. 2000 (in Dutch)
3. Rotmans, Kemp and Van Asselt, 2001

4. See, for example, Sustainable Technological Development - Knowledge Transfer and Embedment, 2000, p. 45
5. See also Anderson, 1995
6. Knowledge Centre PPP, 1998, Van Ham and Koppenjan, 2002, pp.440-441
7. We introduced eight competing values in an earlier publication for the Netherlands Consultative Committee of Sector Councils for Research and Development. De Bruijn and van der Voort, 2000
8. Interview with Peter Folstar
9. Interview with Bertrand van Ee
10. Described in detail in de Bruijn et al, 1998
11. This process is now in progress: www.benchmarking-energy.nl
12. De Bruijn and Ten Heuvelhof, 2000, pp.35

Annex 1

BIBLIOGRAPHY AND REFERENCES

Anderson, J, Public Policymaking, Houghton Miflin, Boston MA, 1997

Army Corps of Engineers & South Florida Water Management District (1999) *Rescuing an Endangered Ecosystem; The Plan to Restore America's Everglades*. Jacksonville/ West Palm Beach, Fla: United States Army Corps of Engineers and South Florida Water Management District

Army Corps of engineers (2003a) Why restore the everglades. www.evergladesplan.org/about/why restore.cfm, access date 9 September 2003.

Army Corps of Engineers (2003b) http://www.evergladesplan.org/about/rest_plan_02.cfm, accessed 11 September 2003.

Army Corps of Engineers (2003c). *Comprehenisve Everglades Restoration Plan Aquifer Storage and Recovery Program* http://www.evergladesplan.org/facts_info/sywtkma.cfm, accessed 15 September 2003.

Army Corps of Engineers (2003d). Cape Sable Seaside Sparrow. http://www.evergladesplan.org/facts_info/sywtkma_sparrow.cfm, accessed, 16 September 2003.

Army Corps of Engineers (2003e) Recover. http://www.evergladesplan.org/pm/recover/recover.cfm, accessed 16 September 2003

Beeching, The Reshaping of British Railways, HMSO, 1963

Brugh, M. aan de, The young vaccine entrepreneurs, in NRC Handelsblad, 10 January 2002 (in Dutch)

Bruijn, J.A. de and E.F. ten Heuvelhof, Networks and Decision Making, Lemma, Utrecht, 2000

Bruijn, J.A. de, E.F. ten Heuvelhof and R.J. in 't Veld, Process management: Decision-making about the environmental and economic aspects of packages for consumer products, Delft, 1998 (in Dutch)

Bruijn, J.A., and H.G. van der Voort, Public-private partnership and Scientific Research; a Framework for Evaluation, Consultative Committee of Sector Councils for Research and Development, The Hague, 2000

Clark, A, Railroader at the halfway house; an interview with Ian McAllister, chairman of Network Rail, in The Guardian, 5 October 2002

Cullen, The Ladbroke Grove Rail Inquiry, 2001

Davis, S. and J. Ogden (eds.), Everglades – The Ecosystem and its Restoration: Delray Beach, florida, St. Lucie-Press, 1994

Dawkins, R, The Selfish Gene, Oxford University Press, Oxford, 1978
DTOTransitions in practice; Sustainable Technological Development – Knowledge Transfer and Embedment, (DTO- KOV), October 2000 (in Dutch)
Eeten, M.J.G. van, and E. Roe: Ecology, Engineering and Management, Reconciling Ecosystem Rehabilitation and Service Reliability, Oxford University Press, 2002
Fukuyama, F, Our Posthuman Future: Consequences of the Biotechnology Revolution, Farrar, Straus & Giroux, New York, 2002
Gibb, R. et al, The Privatization of British Rail, in Applied Geography, vol. 16, no.1, 1996, pp. 35-51
Governor's Commission, Biennal Report, Maintaining the Momentum: In South Florida Ecosystem Restoration Task Force. http://www.sfrestore.org/documents/biennialreport/01.htm Accessed September 11, 2003
Gunderson, L, Resilience, fleitility and adaptive management: antidotes for spurious certitude?, in Conservation Ecoloty 3 (1). http://www.consecol.org/Journal?vol3/is1/art7/index.html Accessed: September 19, 2000, 1999
Ham, H. van and J. Koppenjan (eds), Publiek-private samenwerking bij transportinfrastructuur; Wenkend of wijkend perspectief?, Lemma, Utrecht, 2002 (in Dutch).
Howells, J, Regional Systems of Innovation, in D. Archibugi, J. Howells and J. Michie (eds.), Innovation Policy in a Global Economy, University Press Cambridge UK, 1999
Jackson, B.A., Innovation and Intellectual Property: The Case of Genomic Patenting, in Journal of Policy Analysis and Management, Vol. 22, No. 1, 5-25 (2003)
Kay, J, Privatization in the United Kingdom, 1979-1999, 2001, www.johnkay.com
Kasteren, J. van, System innovations; The long march through the institutes, in the Ingenieur, www.ingenieur.nl 2001 (in Dutch)
Knowledge Centre op Public- Private Partnerships of the Dutch Ministry of Finance, Meer Waarde door Samen Werken, The Hague, 1998 (in Dutch)
Ministry of Economic Affairs, The journey; Transition to a sustainable energy management, December 2001 (in Dutch)
Monami, E, European Passenger Rail Reforms: A Comparative Assessment of the Emerging Models, in Transport Reviews, vol.20, no.1, 2000, pp.91-112
National Council for Agricultural Research, Innovating with ambitions; opportunities for agro business, green space and the fishing industry, report No. 99/17, 1999 (in Dutch)
Office of the Rail Regulator, *Accountability of Railtrack*, May 2001
Office of the Rail Regulator, *Annual Report 2001-2002*, October 2002
Office of the Rail Regulator, *ORR Draft Business Plan 2002-2003*, August 2002
Ogden, J, Everglades- South Florida Assessments, In K. Johnson, F. Swanson, M. Herring and S. Greene (eds.), Bioregional Assessments: Science at the Crossroads of management and Policy, Island Press, Washington DC, 1999, pp. 169-185

Perrow, C, Normal accidents; living with high risk technologies, Princeton, USA, 1999

Preston, J. and A. Root, Great Britain, in D. van de Velde, Changing Trains, Railway Reform and the Role of competition; The Experience from Six Countries, Ashgate, Aldershot, 1999

Ridley, M, Genome, The Autobiography of a Species in 23 Chapters, Fourth Estate, London, 1999

Robbins- Roth, C , From Alchemy to IPO; The Business of Biotechnology, Perseus Publishing, Cambridge MA, 2000

Rotmans, J, R. Kemp, M. van Asselt, F. Geels, G. Verbong and K. Molendijk, *Transitions and Transition management; The case study about low-emission energy supply*, International Centre for Integrative Studies, October 2000 (in Dutch)

Rotmans, J, R. Kemp and M. van Asselt, More evolution than revolution; Transition management in public policy, in *Journal of futures studies, strategic thinking and policy*, vol. 03, No. 01, February 2001

Shaw, J. competition in the UK Passenger Railway Industry: Prospects and Problems, In Transport Reviews, vol.21, no.2, 2001, pp. 195-216

South Florida Water Management District, Governing Board, http://www.sfwmd.gov/gover/2_govboard.html, accessed September 10, 2003

Strategic Rail Authority, Moving forward: Leadership in partnership; Annual Report 2001-02, July 2002

Strategic Rail Authority, *The strategic Plan: Summary*, January 2002

Thompson, L, Changing Railway Structure and Ownership: Is Anything Working? In Transport Reviews, vol.23, no.3, 2003, pp. 311-355

Tomlinson, H, PFI firms in a dark tunnel as Tube deal gets the green light, in *The Independent on Sunday*, 6 October 2002

Vrakking, W, Systcm innovation; Thc stratcgy of thc imagination, in *www.managementsite.nl*, 2000 (in Dutch)

Wolff, G, The Biotech Investors Bible, John Wiley & Sons, New York, 2001

Annex 2

LIST OF RESPONDENTS

Case study: Comprehensive Everglades Restoration Program

Stuart J. Appelbaum
US Army Corps of Engineers
Jacksonville District
Jacksonville, Florida, USA

Steve Davis
South Florida Water Management District
West Palm Beach, Florida, USA

Donald L. DeAngelis
University of Miami
Biological Resources Division - USGS
Department of Biology
Coral Gables, Florida, USA

Lance Gunderson, Ph.D.
EMORY University
Department of Environmental Studies
Atlanta, Georgia, USA

John C. Ogden
South Florida Water Management District
West Palm Beach, Florida, USA

Case study: The Rail Revolution

Graham Eccles
Executive Director
Stagecoach Rail
Virgin Rail Group
London, UK

David Thomas
Director Corporate Finance
Strategic Rail Authority
London, UK

Agnes Bonnet
Head European Affairs
Strategic Rail Authority
London, UK

Andrew Burgess
European Policy and Competition Specialist
Office of the Rail Regulator
London, UK

John Smith
Director of Regulation
Network Rail
London, UK

Anson Jack
Head of Strategy
Network Rail
London, UK

Derek Holt
Senior Managing Consultant Regulation
OXERA (www.oxera.co.uk)
Oxford, UK

Case study: The Boston biobang

Stephen H. Atkinson
Vice-President Commercial Development
Acambis
Boston, Massachusetts, USA

Erik Kuja
Analyst Strategic Alliances
Pfizer Development Technology Center
Boston, Massachusetts, USA

Stephen Mulloney
Director Government Regulations and Communication
Massachusetts Biotechnology Council (MBC)
Boston, Massachusetts, USA

Una Ryan, PhD
President and CEO
Avant Immunotherapeutics, Inc.
Needham, Massachusetts, USA

Jonathan Seals PhD
Director Process Research and Development
Biochem/ Shire
Northborough, Massachusetts, USA

Paul Wengender
Research Scientist Protein Sciences
Pfizer Development Technology Center
Boston, Massachusetts, USA

Test interviews

Bertrand van Ee
Vice-president Infrastructure
Fluor Daniel BV

Prof.dr.ir. Peter Folstar
Initiative Director Genomics
NWO (Netherlands Organisation for Scientific Research)

Dr Gertjan Fonk
Green Space and Agrocluster Innovation Network

Dr Henk van Latesteijn
Head of Strategic Policy Making Bureau
Ministry of Agriculture, Nature Management and Fisheries

Dr Karel Mulder
Associate Professor of Technology Assessment
Delft University of Technology

Annex 3

MEMBERS OF THE ADVISORY COMMITTEE

Prof. H.O. Voorma Ph.D.
Chair of the Consultative Committee of the Sector Councils (COS)

Dr. G. Vos
Director Dutch National Council for Agricultural Research / Netherlands Green Space and Agrocluster Innovation Network

S. Eschen MPA, M. Sc., Dutch Ministry of Justice, member for the preparation of the Public Management, Justice and Safety Sector Council

J. H. of der Veen, M. Sc.
Director Netherlands Study Center for Technology Trends

P. Morin, M.A.
Secretary of the Consultative Committee of Sector Councils for Research and Development

Annex 4

THE DUTCH CONSULTATIVE COMMITTEE OF SECTOR COUNCILS FOR RESEARCH AND DEVELOPMENT

The Consultative Committee of Sector Councils for research and development (COS) in the Nethetlands is the collaboration platform for the system of cooperating sectorcouncils and other COS-members active in foresight. The basis is formed by a framework law.
A sectorcouncil comprises respresentatives from the scientific community, society and industry and the government. Together they present an independent vision of the knowledge-needs and the priorities for strategic research in their respective sectors. Their vision is based on thorough studies, such as foresight activities, necessary to obtain a long-term perspective on societal and scientific trends. From a viewpoint of a necessary integral approach sectorcouncils are very often working together so as to consider one trend in coherence with another. Sectorcouncils also map out developments in science and technology and the implications for society.
Area's covered by the system are presently: Spatial planning, nature and Environment, rural areas and agriculture, health, technology, development assistance and (summer 2004) public administration, justice and security and education. The possibility of a sector council for transport and infrastructure is under investigation.
Functions of the COS as a collaboration platform are e.g. promoting a joint approach in foresight- and programming studies and studies on the development of methodology, funded by the COS Coordination Fund. Furthermore the COS sees to joint input in administrative consultations with ministeries and other organisations. For more information please visit www.minocw.nl/cos

COS- members are:

- Innovation Network Green Space and Agriculture
- Netherlands Development Assistance Research Council (RAWOO)
- Advisory Council for Research on Nature and the Environment (RMNO)
- Netherlands Study Center for Technology Trends (STT)
- Advisory Council on Health Research

For the following sectors a sector council is in preparation:

- Public Administration, Justice and Safety (2004)
- Education
- Traffic and Infrastructure
- Labour
- More information on the COS-members is available on the follow-ing website: www.minocw.nl/cos

For Product Safety Concerns and Information please contact our EU representative GPSR@taylorandfrancis.com Taylor & Francis Verlag GmbH, Kaufingerstraße 24, 80331 München, Germany

T - #0038 - 230425 - C0 - 254/178/6 [8] - CB - 9789058096722 - Gloss Lamination